BIOLOGY LABORATORY MANUAL

Third Edition

MARJORIE S. SHARP
University of Texas at Arlington
Arlington, Texas

Hunter Textbooks Inc.

ISBN 0-88725-009-2

Inquiries should be addressed to the publisher:

823 Reynolda Road
Winston-Salem, North Carolina 27104

TABLE OF CONTENTS

INTRODUCTION

GENERAL INSTRUCTIONS

1. Be on time so you won't miss any little "pearls of information" your instructor may drop. Opening remarks will cover material to be included on exams.
2. Participation in the laboratory and satisfactory completion of all material will be required of each student — or they will be shot at sunrise — or worse still, not receive credit for the course.
3. Please do not smoke or bring food or drinks into the lab. (Pickled worms and snails don't go well with hamburgers!)
4. If absent, you are still responsible for all work covered in that period. Therefore, an absence should be made up in a later lab that same week. Because of space limitations there is no guarantee that we can put you in another lab. If at all possible, try to attend another section taught by your instructor. Lab absences count two hours and excessive absences can result in an automatic drop and an "F" grade.
5. Textbook and lab manual will be needed each lab period.
6. PLEASE put away all supplies and equipment and leave tables and working areas clean. Thanks!

WHAT CAN YOU DO TO GET THE MOST OUT OF LAB?

If the only reason for learning about the biological world involved passing an exam, there would be no need for a lab. Facts can be obtained from books. Observations can be more meaningful (and retained far longer) than can rote memory of facts.

Your instructor is here to help guide you, but will be anxious to assist you only after you have first tried to help yourself. In the final analysis — you are the one who must do the work in order to understand the materials being presented — your instructor can only assist in the process.

Suggested steps to follow:

1. Since laboratory time is always inadequate, prepare in advance by reading the exercise and pertinent sections in your text.
2. Most studies are individual in nature and you should keep them so. Do not permit other students to copy your drawings or answers. Why do their work? Consulting with other students is a good idea, however. They may see something you don't.
3. **Think** about the exercise. Don't just go through and blindly fill in blanks and look at slides. Try never to carry out a step without knowing why you are doing it. A laboratory is a unique opportunity to learn by observation; don't waste it.
4. It may be helpful to make a brief outline (before or after doing the work). If done before it can prevent frenzied and aimless activity in a lab which may be a bustling and, at times, noisy place. After doing the work you can then make this outline more detailed and it could prove useful for study purposes.
5. Be considerate of others. If a procedure demands team effort, do your share. If it requires individual work, do your own and let others do theirs in peace and quiet.
6. Biology is not a disorganized mass of unfamiliar terms, despite what you may think. In these labs you will be introduced to new terminology (significant terms are usually in bold type), but concepts and broad principles relating to life activity are of prime importance. Concentration on overall principles in each session will allow specialized terms to fall into place.
7. Place in your manual neat notes, drawings, labels and answers to questions. These are not done for the sake of edification of the instructor or for a grade. Clear, concise records are a means of systematizing and improving your learning. The manual should be filled in during the period **as you do the work.** Filling it in ahead of time fills up the blanks on the paper but does little to help you fill in the "blanks" in your brain.

EXAMS AND GRADING

1. The laboratory grade will be based on the following.

   2 practical exams = 50%
   daily exam average = 50%

   or

   3 practical exams = 100%

   Exams are so designed (or at least that's what we've tried to do!) that they should not be difficult if YOU have done the work yourself.

2. The following will also be considered:
   a. Ability to apply knowledge gained (as evidenced by answers to questions and participation in discussions).
   b. Ability to work independently and to do the work assigned.
   c. Ability to record information (drawings and notes in the manual).
   d. Attendance and attitude. Unexplained absences can result in loss of points from the final average. The **student** is responsible for reporting reasons for all absences to his or her lab instructor. Exams missed and not made up in another section will be recorded as a zero.

3. The overall course grade will be computed as follows:

Laboratory grade		= 25%
Lecture	Three one-hour exams	= 50%
	Final exam	= 25%

LABORATORY 1

SCIENTIFIC MEASUREMENT, THE SCIENTIFIC METHOD AND THE CHEMICAL NATURE OF LIFE

MEASUREMENT AND THE METRIC SYSTEM

OBJECTIVES: To become familiar with:
1. the concept of measurement
2. the metric system
3. the orderly recording of data

Science is concerned with discovering the mechanisms which operate in the physical world. Consequently, it is often necessary to measure certain physical properties in some standard way. The more commonly measured characteristics of objects include:

1. **size** — length, width, and depth
2. **volume** — the amount of space occupied by an object
3. **mass** — the quantity of matter present in an object; its bulk
4. **weight** — the gravitational pull on an object's mass
   The student should note that the mass of a body remains constant, but its weight varies with changes in gravity. (The same object weighs very slightly less on top of Mt. Everest than it does at sea level.)
5. **temperature** — the amount of heat present in a body
6. **time** — the duration of an event

An important quality in a laboratory investigator is the ability to make accurate measurements. Not all measurements can or need to be made with the same degree of accuracy. A geologist may only be able to estimate the age of a rock within several hundred years, while measurements of a fraction of a second are crucial in determining the speed of light. Generally, the degree of accuracy desired should be in reasonable agreement with the quantity being measured.

Within the scientific community and throughout much of the world, the metric system of measurement is generally preferred over the English system of yards, feet, etc., because of the greater ease of conversion between units.

The basic unit of length is the **meter** (39.37 inches or slightly more than one yard). The basic unit of volume is the **liter** (1.1 quarts). The basic unit of mass is the **gram:** gravity exerts a force of one pound on a mass of 454 grams at sea level.

Each of these units may be combined with a prefix to indicate either larger or smaller units.

deci (d) — (one tenth of)
centi (c) — (one hundredth of)
milli (m) — (one thousandth of)
micro (μ) — (one millionth of) One micrometer is called a micron.
kilo (k) — (one thousand times)

Thus 1000 mm = 100 cm = 1 m = 0.001 km.

Conversions between the units are made by multiplying or dividing by powers of ten, or by simply moving the decimal point to the right or the left. This is the same type of manipulation involved in changing pennies into dollars, since our monetary system is also based on powers of ten. As an example, consider changing 250 ml to liters. Since there are 1000 ml in 1 liter, the conversion is made by dividing by 1000.

$$250 \not{ml} \times \frac{1 \text{ liter}}{1000 \not{ml}} = \frac{250 \text{ liters}}{1000} = 0.25 \text{ liters}$$

Note that this is equivalent to simply moving the decimal point three places to the left.

Three temperature scales are commonly used to measure the intensity of heat possessed by an object or its environment. The **Fahrenheit** temperature scale is in common household use in the United States. Most scientific work uses either the **centigrade (Celsius)** or the **absolute (Kelvin)** temperature scale. The centigrade scale (which will be employed in this laboratory) and the Fahrenheit scales are compared in **Figure 1.1.** Conversions from one temperature scale to the other may be made using the following formulas which were derived by comparing the boiling and freezing points of water on each of the temperature scales.

$$\text{temperature in °C} = 5/9 \times (\text{temperature in °F} - 32°)$$
$$\text{temperature in °F} = 9/5 \times (\text{temperature in °C}) + 32°$$

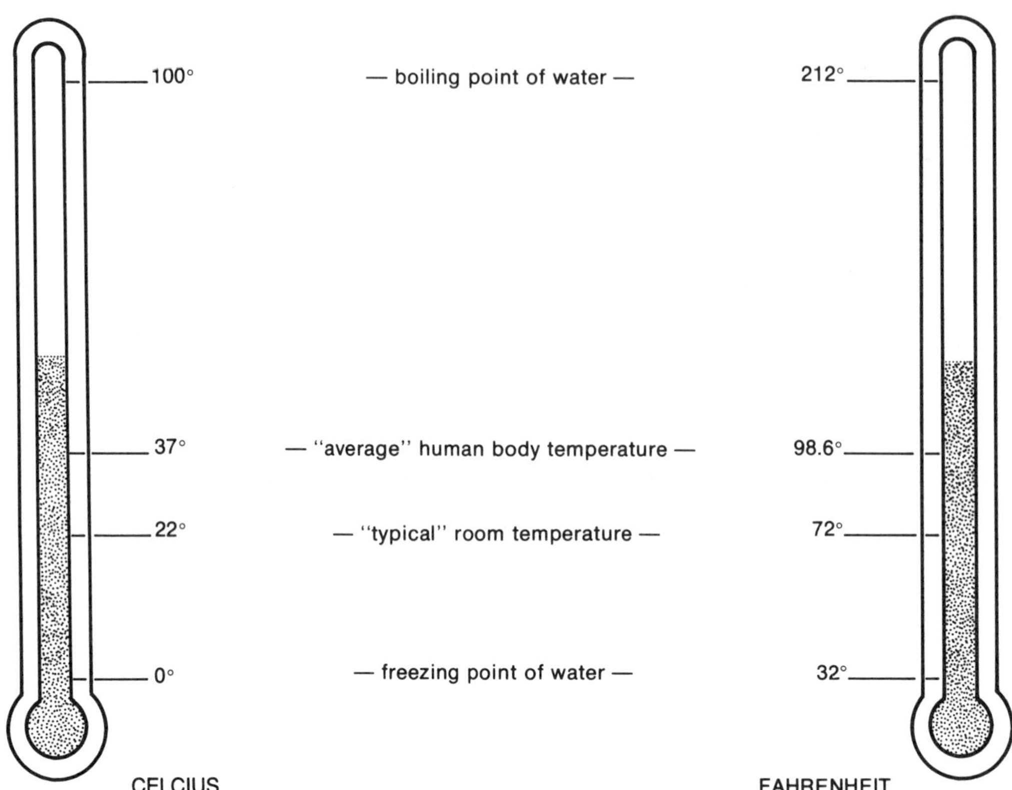

Figure 1.1. Comparison of celcius and Fahrenheit temperature scales.

1. Each student should individually measure in meters the length and width of the lab table. Record these and convert your measurements into the other units listed.

Lab Table	**Width**	**Length**
meters	__________	__________
centimeters	__________	__________
millimeters	__________	__________
kilometers	__________	__________

2. Working in pairs, use a 15 cm plastic ruler to measure the diameter and length (to the nearest mm) of a cylindrical piece of potato. Record this data in the space provided.

3. Using the triple-beam balance, weigh the potato core to the nearest 0.1 gram. **Dry the core first** to remove excess moisture. Record data.

4. Determine the volume of the potato core to the nearest ml by measuring the volume of water it displaces. To do this, fill a graduated cylinder to the 85 ml mark. Stick a teasing needle in the top of the core and lower the core into the cylinder until the top is submerged. Read the new water level and determine the core's volume by subtracting 85 from the higher number. Note: When water is placed in a cylinder it tends to adhere to the cylinder's wall, causing a concave liquid surface called a **meniscus.** The correct reading is made by selecting the mark nearest the bottom of the meniscus **(Figure 1.2).**

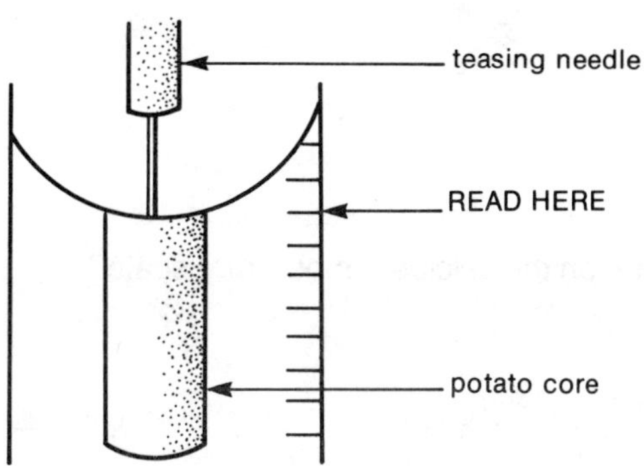

Figure 1.2. Schematic diagram illustrating how to read meniscus level.

5. Place the core in a beaker containing 25 ml of 20% salt solution. Record the time. Allow it to soak until you have finished the second part of the lab exercise (at least 30 minutes). Then remove the core, note the time, and repeat steps 2, 3, and 4. **Dry the core** before reweighing. In the space provided describe any change in appearance or consistency of the potato which has occurred. **Return salt solution to stock container and potato core to distilled water.**

Potato Core	**Before soaking**	**After soaking**
diameter (mm)	12	12 (even)
length (mm)	50	50 (even)
volume (ml)	4	4 (even)
weight (g)	5.2	5.4 (+2)

Time soaking began 2:30 Time soaking ended 3:13 Minutes soaked 43

Appearance of potato – mushy

After the potato was soaked, which measured quantities changed? What caused these changes?

Weight (absorption)

What would you expect to happen if you soaked the potato core twice as long? Four times as long?

Weight ↑ 2x
4x

QUESTIONS:

1. There are 453.59 grams in one pound. How many kilograms are there in 3 pounds?

2. What is the freezing point of water on the celcius temperature scale?

3. Using the only available thermometer, you find your body temperature to be 38.0° C. What would your temperature be if you measured it using a Fahrenheit thermometer?

4. Convert −40° Fahrenheit to the equivalent celcius temperature.

5. Which represents a greater amount of heat energy — a temperature change of 1° F or 1° C?

THE SCIENTIFIC METHOD AND THE CHEMICAL NATURE OF LIFE

OBJECTIVES: 1. To illustrate the use of the scientific method of investigation.
2. To discover the chemical nature of some biological materials.

A scientist is frequently interested in establishing a cause-and-effect relationship between two events, so that the nature of such occurrences may be predicted in the future. Additionally, as you will do, he may seek to analyze natural phenomena by measuring or testing some selected property. There are certain fundamental procedures followed by investigators in all fields of science when attempting to discover scientific information. Standard procedures are necessary to insure that the information obtained actually represents physical reality, and that similar data can be obtained by other investigators. Normally, somewhere in the course of experimentation, the following will occur.

1. **Observation** — noting a phenomenon, event, or sequence of events, and recording the observations.
2. **Problem** — stating as precisely as possible what one is trying to determine. This is conveniently done by asking a relevant, answerable question.
3. **Hypothesis** — a tentative answer to the previously stated question, based on available information (observations). Since the observation will likely permit more than one answer, any hypothesis offered as an explanation must be testable.
4. **Experimentation** — one or more tests designed to confirm or refute the hypothesis. Using a two-part test, the experimenter seeks to eliminate all but one possible answer to the question asked. Only one characteristic or condition should be varied at a time, and the effects of its variation should be compared with a similar test in which it is kept constant. The group of specimens or series of tests in which conditions are varied is known as the **experimental,** while the unvaried group is the **control.** The condition which differs in the two groups is called the **variable.**
5. **Conclusions** — If experimental results confirm the hypothesis, one may conclude that the hypothesis is correct. If the experimental results conflict with the hypothesis, it must either be abandoned or modified to take into account the new information.

You will test for the presence of certain specific types of chemicals in four materials of living origin. This should give some insight into the "living substance" of which cells are made. The materials to be tested are milk, heart, potato and soybeans. The tests will also be run on distilled water as a control.

Test Name	Tests For	Positive Result	Negative Result
1. Biuret	Protein	Violet color	Pale blue color
2. Iodine	Plant Starch (Carbohydrate)	Purplish-black color	Red-brown color
3. Sudan III	Lipids (Fatty substances)	Red dye powder dissolves	Red dye powder floats on surface

The tests you will be performing are qualitative — they indicate only whether a substance is present or absent. You may be able to tell if the chemical is present in very large or very small amounts by the degree of color change, but it would require additional equipment and/or chemical tests to quantitatively determine exact concentrations.

METHOD

Work as a group of four. Label each of the fifteen clean test tubes as 1a, 1b, 1c, 2a, etc., to 5a, 5b, and 5c. Record your results as a positive (+) or negative (–) test for each in the space provided.

Make certain that test tubes are clean before you begin.

1. Add twelve drops of distilled water (not tap water) to1a, 1b, and 1c. Then add the following test chemicals to the appropriate tubes in series 1. These 3 tubes will constitute a set of negative controls.
   a. **Biuret test.** Add twelve drops of sodium hydroxide (NaOH) solution to tube a. (CAUTION: NaOH will burn. Handle with care. If spilled, wash your skin and/or clothing in cold running water immediately. Wipe away thoroughly any spillage in work area.) Add five drops copper sulfate ($CuSO_4$) solution. Gently shake. Observe any color change, and record test results. **Save the tube for comparison with tests to be run later.**
   b. **Iodine test.** Add five drops of iodine solution to tube b. Gently shake. Observe any color change and record. **Save the tube** for comparison.
   c. **Sudan III.** Add a small amount of Sudan III to tube c by scooping up some of the powdered dye on the end of a toothpick. Gently shake. Observe what happens to the dye and record. **Save for comparison.**
2. Add twelve drops of milk, using a **clean dropper,** to tubes 2a, 2b, and 2c. Repeat steps a, b, and c from above, comparing with the appropriate control.
3. Add twelve drops of heart extract, using a clean dropper, to tubes 3a, 3b, and 3c. Repeat steps a, b, and c, comparing with the appropriate control.
4. Add twelve drops of potato extract to test tubes 4a, 4b, and 4c, using a clean dropper. Repeat as above.
5. Add twelve drops of soybean extract to the test tubes 5a, 5b, and 5c, using a clean dropper. Repeat as above.

It may be necessary to allow the tests to sit a few minutes to give sufficient time for all chemical reactions to occur. When you have completed the tests and recorded the results, confirm your conclusions with the instructor. Then **clean all test tubes thoroughly and invert them in the rack.**

RESULTS

Test	**1. Control**	**2. Milk**	**3. Heart**	**4. Potato**	**5. Soybean**
Biuret (protein)	________	________	________	________	________
Iodine (plant starch)	________	________	________	________	________
Sudan III (lipid)	________	________	________	________	________

QUESTIONS

1. State the hypothesis behind the test using tube 3a.

2. Can this hypothesis be proven using only tube 3a? Why or why not?

3. Considering together tubes 1b and 4b, which is the control? ______________________

Which is the experimental? ______________________

What is the variable? ______________________

4. If you obtained negative results in all tests for protein you performed using the Biuret test reagents, would this prove that protein was not present in the test materials? Why? What additional type of chemical might you wish to test?

5. In 1824, the name protoplasm was suggested for the substance making up living cells. Protoplasm was originally thought to consist of only a single, uniform chemical substance. Do your results support this hypothesis?

6. Propose an alternative to the hypothesis in question 5.

7. Must the results of an accurate control test always be negative? Why or why not?

LABORATORY 2

THE CHEMICAL STRUCTURE OF LIVING ORGANISMS

OBJECTIVES: To become familiar with:

1. the basic chemical structure of biological molecules
2. types of bonds between molecules and atoms
3. the relationship of molecular structure to function

To conserve lab time, paper models on pages 21-25 should be cut apart prior to coming to class. Separate the components by cutting on the **solid lines** and store in the envelope just inside the back cover of the manual.

ATOMIC STRUCTURE

Living organisms are composed of large organic or carbon-containing molecules which are principally made up of four different kinds of atoms — carbon, nitrogen, hydrogen and oxygen. The earth is composed of 106 known chemical **elements,** or types of **atoms.** Atoms are composed of three different particles with the following characteristics.

Particle	Mass	Electrical Charge
proton	1	positive
neutron	1	neutral
electron	negligible	negative

An element's **atomic number** equals the number of its protons. Carbon, for example, has six, while oxygen contains eight. Protons and neutrons are concentrated near the atom's center (nucleus). The number of neutrons may vary from atom to atom. Atoms of a single substance containing differing numbers of neutrons are known as **isotopes.** The sum of the number of protons and average number of neutrons (based on the abundance of the various isotopes) determines the mass of a particular element or its **atomic weight.**

Proton and electron numbers in an atom are equal, therefore the atom is electrically neutral. The negative electrons orbit the positive nucleus in specific electron shells, each capable of holding some maximum number of electrons. In one scheme for illustrating this, the first shell holds only two, while the second may contain up to eight. Generally, one electron shell is completed before the next is begun.

1. A **periodic table** of the elements is available in the lab. Find sodium (chemical symbol Na) on the chart and complete **Figure 2.1.**
2. Use the periodic table to complete the following.

Element	Chemical Symbol	Atomic Number	Atomic Weight	No. of Outer Shell Electrons
carbon	C	______	______	______
hydrogen	H	______	______	______
nitrogen	N	______	______	______
______	Cl	______	______	______
______	Ca	______	______	______
potassium	______	______	______	______

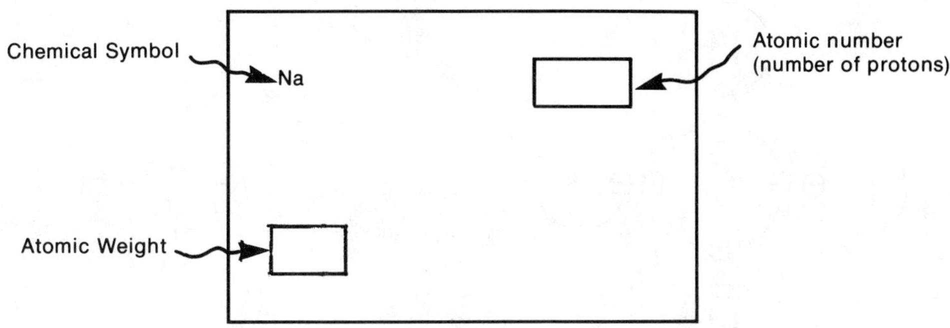

Figure 2.1. Periodic table information for sodium.

3. On the average, how many neutrons are found in each of the following? Give answer to the nearest whole neutron.
(Atomic weight — Number of protons = Number of neutrons)

carbon________ potassium________ sulfur________ oxygen________

BONDING

Atoms are generally most chemically stable when their outer shell contains its maximum number of electrons or contains eight electrons, whichever comes first. Many chemical reactions involve redistribution (sharing or actual transfer) of the outer shell electrons of two or more atoms to achieve this stability.

IONIC BONDS

Electrically charged atoms are known as **ions,** and the attraction between oppositely charged ions forms **ionic chemical bonds.** Ionic bonds are most common with atoms which must gain or lose only one or two electrons to complete their outer shells. Under proper conditions, a sodium atom's one electron may be transferred to a chlorine atom (which has seven electrons), resulting in stable electron distributions in both. The sodium is now a positively charged sodium ion, having lost one electron. The chlorine has acquired an electron and is negatively charged.

COVALENT BONDS

Covalent chemical bonds result when electrons are shared by two or more atoms. These bonds usually form between two atoms with an intermediate number of outer shell electrons. The outer shell of carbon has four electrons which it can share with another atom. Hydrogen has only one electron. How many pairs can it share? _______

The covalently bonded compound methane results when a carbon atom (C) shares an electron pair with each of four hydrogen (H) atoms **(Figure 2.2A).** Each electron of the shared pair orbits both the carbon and hydrogen atoms. By convention each shared electron pair is represented by a single line **(Figure 2.2B).** The sharing of a single pair of electrons by two atoms produces a single covalent bond. If these atoms share two electron pairs, a **double bond** is formed, and if they share three pairs, a **triple bond** results. Through some combination of single, double, or triple bonds with one or more atoms, carbon generally shares four electron pairs, hydrogen shares one, and oxygen shares two electron pairs.

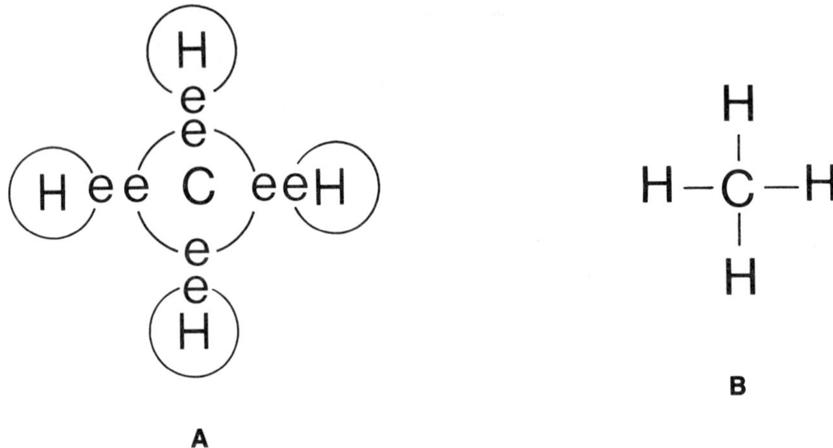

Figure 2.2. Covalently bonded methane: **A,** shared electron pairs; **B,** conventional representation of electron pairs.

OXIDATION-REDUCTION

A chemical process which involves loss of electrons by an atom is called **oxidation,** and the substance losing the electrons is said to have been oxidized. A reaction in which electrons are gained by a substance is termed **reduction,** and the substance gaining the electrons has been reduced. Ionic reactions are often termed oxidation-reduction reactions, since they involve both electron loss and gain. Although covalent reactions do not involve actual electron transfer, the sharing of the electron pair is often unequal and may be thought of as partial electron loss. This occurs when a small atom such as hydrogen is sharing electrons with a considerably larger one, such as oxygen. In living organisms, partial electron loss (oxidation) or gain (reduction) is accompanied by the addition or the removal of hydrogen or oxygen from the molecule. The following summarizes these aspects of oxidation-reduction.

OXIDATION: electron loss
often accompanied by

1. addition of oxygen, and/or
2. removal of hydrogen

REDUCTION: electron gain
often accompanied by

1. addition of hydrogen, and/or
2. removal of oxygen

Will the element calcium tend to form an ionic or a covalent bond?________________

Why?________________________________

Chlorine?________________________________

Based on their electron structure, in what ratio would they combine?________________

Which atom is oxidized?________________Why?________________

Which is reduced?________________________________

What kind of chemical bonds will carbon tend to form?________________

In what ratio will it combine with oxygen?________________

How many electron pairs are shared by each carbon and oxygen atom?________________

Draw diagrams similar to **Figure 2.2** representing the stable compound formed by the combination of carbon and oxygen.

THREE-DIMENSIONAL CHEMICAL STRUCTURE

Atoms and chemical bonds have thus far been represented as two-dimensional structures. Such representations are useful and may often be necessary to maintain clarity in diagramming large molecular structures. As matter is three-dimensional, however, it is advantageous to gain some insight into spatial patterns of organic molecules.

Atoms will be represented by the colored balls contained in the molecular models kit. Each color represents a different atom as follows:

carbon — black
hydrogen — white
nitrogen — orange
oxygen — red
chlorine, bromine, iodine — green
phosphorus — purple
sulfur — yellow

The number of holes in each model atom reflects the number of bonds it usually forms. Each spring represents a single shared pair of electrons, or chemical bond. The longer springs should be used to represent bonds between the following:

C — C C = O (double bond = 2 springs) C — N

The shorter springs should be used for bonds between:

C — H C — O (single bond) O — H N — H

Bond length differences correspond approximately to known atomic spacing. Assemble the models by inserting the springs into the holes with a **firm, clockwise** twist. Disassemble using the same clockwise turn in order to prevent damage to the spring.

WATER

1. **Construct a model of a water molecule.** The H atoms are connected to the O atom by single bonds (short springs).

2. **Remove one of the white balls from the molecule.** The white ball removed represents a positively charged **hydrogen ion** (H^+) **(Figure 2.3).** The negatively charged portion of the model is an **hydroxyl ion** (OH^-). Such a separation into ions occurs to a limited extent in water and forms the basis for determining the degree of acidity or alkalinity of a solution. What does the spring attached to the oxygen (red ball) represent? ______________________________

Why is the hydrogen ion positively charged? ______________________________

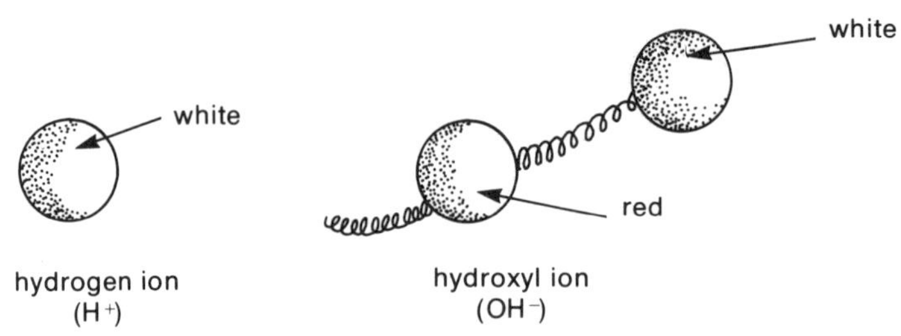

Figure 2.3. Hydrogen ion and hydroxyl ion models.

OXIDATION-REDUCTION OF SOME ORGANIC MOLECULES

1. **Draw the structural formula of a molecule of methane** (CH_4), a component of natural gas. **Construct a model of the CH_4 molecule.**

Is the carbon reduced or oxidized in this molecule? ______________________

Why? ______________________

2. **Substitute an OH group for one of the hydrogens.** This new molecule is methyl alcohol. The CH_3 group is known as a **methyl** group, the OH as a **hydroxyl** group. All alcohols contain a hydroxyl group.

3. **Write the chemical and structural formulas for methyl alcohol.**

chemical formula | structural formula

4. **Remove two more of the hydrogens and substitute a single oxygen** in their place, using a double bond. This molecule is formic acid. All organic acids contain a **carboxyl** (COOH) group. Is this molecule more oxidized or more reduced than methane? ______________________

5. **Save this model** for use later in the exercise when proteins are discussed.
6. **Draw the structural formula of formic acid** (HCOOH).

7. If the two remaining hydrogen atoms were removed and the second oxygen atom connected to the carbon with a double bond, carbon dioxide (CO_2 or $O = C = O$) would be formed.

8. This series of steps illustrates the oxidation of organic molecules through the addition of oxygen or the removal of hydrogen. If the steps were reversed (going from CO_2 to methane or CH_4), reduction would be illustrated.

BIOLOGICAL MOLECULES

Biological molecules are classified into four major categories: (1) **carbohydrates,** (2) **proteins,** (3) **lipids,** and (4) **nucleotides** and **nucleic acids.** Each grouping contains large molecules which appear quite complex, but which consist of a pattern of smaller, repeating units. Molecules composed of recurring components are called **polymers,** and the subunits are **monomers.** Several different kinds of monomer units may occur within a particular molecule. Thus, an almost infinite variety of distinctly different biological molecules can be formed, based on the number and arrangement of each monomer type present.

CARBOHYDRATES

Simple Sugars

In molecules containing more than two carbon atoms, it is possible to have more than one distinctly different three-dimensional arrangement of the same number and kinds of atoms. One very important group of organic molecules, the **sugars,** will illustrate this.

There are eleven different sugar molecules with the chemical formula $C_6H_{12}O_6$. The structural formulas for two of them are shown in **Figure 2.4.** Only the arrangement of the various atoms differs in the two molecules, but their chemical behavior in living organisms is quite different.

1. **Observe the demonstration models of the glucose molecule,** another $C_6H_{12}O_6$ sugar. The hydrogen and hydroxyl groups must be placed to the left or right of the model exactly as shown in order for the model to represent glucose. **Note the zig-zag pattern** formed by the bond angles between the carbons.
2. When dissolved in water, as in the living cell, glucose occurs principally in the form of a ring, rather than as a straight chain. The straight chain form of glucose can be converted to the ring form by removing the OH group from carbon 5. Then one of the springs of the double-bonded oxygen on C 1 is removed and the oxygen is connected to the vacant hole on C 5. The OH group is inserted into the vacant hole on C 1. See the resulting structural formula in **Figure 2.5** and the display models which illustrate this step-wise sequence.

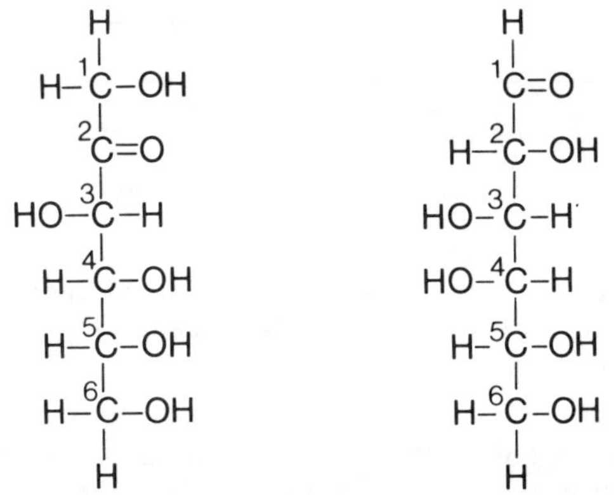

Fructose

Galactose

Figure 2.4. Straight chain forms of some hexoses.

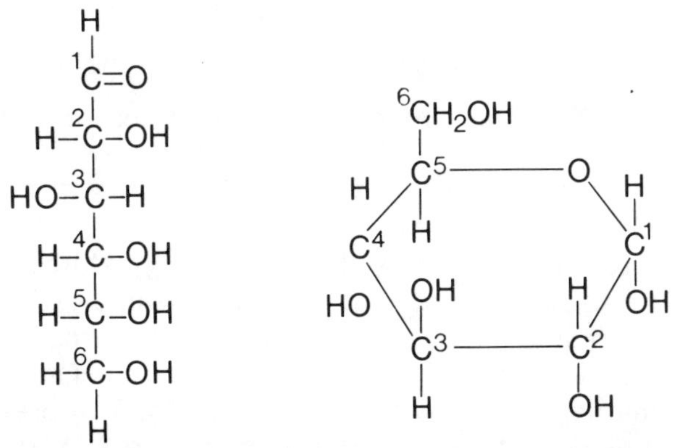

Straight chain form

Ring form

Figure 2.5. Straight chain and ring forms of glucose.

Disaccharides

Simple sugars are the basic monomer units of larger carbohydrate molecules. Single glucose rings may be combined to form larger carbohydrate molecules, such as **disaccharides** (double sugars). One of these disaccharides, formed by the chemical combination of one unit of glucose and one of fructose, is sucrose — ordinary table sugar. The chemical process by which monomer units are linked is **polymerization** or **synthesis (Figure 2.6).**

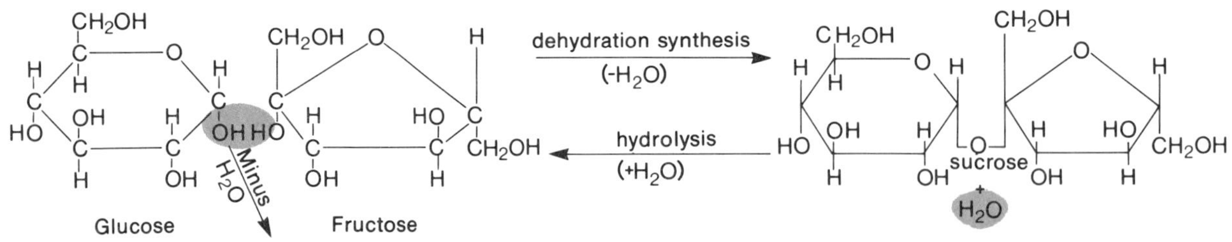

Figure 2.6. Synthesis and hydrolysis of the disaccharide sucrose.

Polysaccharides

Larger carbohydrate molecules composed of three to several hundred sugar monomers are called **polysaccharides. Plant starch,** which is composed of glucose monomer units, is an example. **Glycogen,** another polysaccharide of glucose, is often called animal starch. **Cellulose,** the supporting structure of plant cell walls is also a polysaccharide of glucose. These polysaccharides differ only in the way in which the glucose molecules are bonded together.

1. Two examples of monosaccharides are ______________________________

and ______________________________.

2. The disaccharide sucrose consists of one each of the monomers

______________________________ and ______________________________.

3. Starch, glycogen and cellulose are ______________________________

composed of large numbers of the monomer ______________________________.

4. The combining of two biological molecules is usually accomplished by the removal of

a ____________________ ion and an ____________________ ion and the combining of these two

ions to form a molecule of ____________________.

5. To release the chemical energy stored in larger carbohydrate molecules, living organisms oxidize the component glucose units into carbon dioxide and water. The polysaccharides are broken apart into monomer units by the breakdown process called **digestion.** The links between the monomer units are broken by adding the OH^- and H^+ components of a water molecule to complete the separated portions of the chain. The breaking of a chemical linkage by addition of a water molecule is called **hydrolysis (Figure 2.6).** Hydrolysis occurs in living organisms only under the influence of specific enzymes, protein molecules which will be studied in more detail in the next lab.

Lipids

Lipids, fatty molecules which are largely insoluble in water, perform several functions in organisms. They are important as a major component of cell membranes and are involved in energy storage.

Some of the more simple fat molecules are composed of two types of sub-units — glycerol and fatty acids. Glycerol is a small molecule, containing only three carbon atoms. Attached to the glycerol are three fatty acid molecules. Each fatty acid consists of a long chain of carbon and hydrogen atoms. Although the composition of lipids in nature is quite varied, you will construct only one representative molecule using three of the more common fatty acids — palmitic, stearic, and linoleic acids.

1. **Cut the glycerol and fatty acid molecules** (linoleic, stearic and palmitic) **on the dotted lines of the paper models** (page 23).
2. **Bond together the three water molecules.**
3. **Construct a lipid molecule by matching the tabs on the glycerol and fatty acids.** (Bond palmitic acid to the upper position, linoleic acid to the center position, and stearic acid in the lower position.) The actual metabolic reactions synthesizing lipids are complex, and would require several different enzymes. Just as polysaccharides must be broken into their component units before the stored energy can be used, fats also must be hydrolyzed or digested by specific fat-digesting enzymes.
4. **Acting as an enzyme, break the links between the glycerol and each of the fatty acids.**
5. **Split a water molecule into its ions and complete the component molecules by adding an ionized water to each linkage site.** Did you hydrolyze or synthesize the lipid molecule in steps 4 and 5? ______________________________

PROTEINS

Amino Acid Structure

1. **Using the previously constructed formic acid molecule, remove the hydrogen atom attached to the carbon and substitute a methyl group** (CH_3).
2. **Now remove a hydrogen from the methyl group and substitute an amino group** (NH_2).
3. This molecule is glycine (NH_2CH_2COOH), the simplest of the amino acids. Amino acids are the building blocks of the large protein molecules so important to living organisms. Save this model for later reference. Draw the structural formula of glycine.

4. All amino acids have certain common structural characteristics. **Compare the paper model of the amino acid glycine** (page 27) **with the three-dimensional model previously constructed. Now compare the paper alanine model with glycine.** Note that only the atoms enclosed in the shaded area differ. Each kind of amino acid contains a different grouping of atoms in this position. Locate this area of your three-dimensional model.
5. Proteins or polypeptides are large biological polymers composed of varying sequences and combinations of twenty different **amino acids.** The specific kinds and arrangement of the units determine the way the protein folds into a three-dimensional structure, and thus also specifies its function. Enzymes, which are protein molecules, must have a three-dimensional shape complementary to their substrates. Proteins are also important structural components.

Polypeptide Structure

Proteins are synthesized by **peptide bond** formation between the nitrogen of an amino group of one amino acid and the carbon in the carboxyl group of the next amino acid **(Figure 2.7).** Each bond results in the formation of a molecule of water. Linkage of two amino acids forms a dipeptide. A chain of amino acids formed by peptide bonds is called a **polypeptide.**

1 **Using the paper models, function as an enzyme by forming a peptide bond between the amino acids glycine and alanine.** This will require cutting along the dotted lines on each model. **A water molecule will be formed.**
2. **Bond the amino acid cysteine to the alanine, again forming a water molecule in the process.**
3. Proteins are digested in a manner similar to both carbohydrates and lipids. Specific digestive enzymes hydrolyze the peptide linkages between the amino acid monomers.
4. Since protein structure is determined by amino acid sequence, the polypeptide you formed had a particular three-dimensional structure. The arrangement was glycine — alanine — cysteine. List the different sequences possible using these three amino acids.

   a.______________________ d.______________________

   b.______________________ e.______________________

   c.______________________ f.______________________
5. If a protein molecule contained hundreds of amino acids, there would be thousands of possible arrangements, each resulting in a slightly different three-dimensional structure.

Glycine + Alanine —dehydration synthesis ($-H_2O$)→ Glycylalanine (Peptide link) + H_2O Water

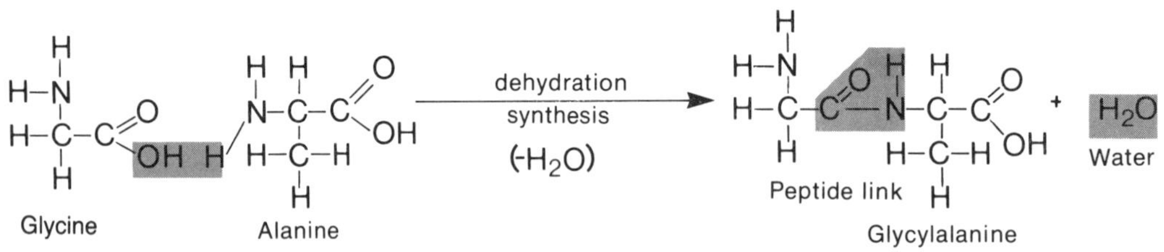

Figure 2.7. Peptide bond formation.

Nucleotides and Nucleic Acids

Nucleic acids, the genetic material of living organisms, direct the process of protein synthesis. Their monomer units are **nucleotides.** Nucleic acid structure will be studied in a later laboratory exercise.

In addition, specific nucleotide molecules act as energy carriers in living organisms. The major energy-carrying molecule is **adenosine triphosphate,** or **ATP.** Three subunits combine to form ATP: a nitrogen and carbon ring structure **(adenine),** a five-carbon sugar **(ribose),** and three **phosphate** groups. Removal of the last phosphate group releases energy and leaves the molecule **adenosine diphosphate (ADP).** The addition of a phosphate group to ADP to reform ATP requires energy, generally supplied by the oxidation of glucose units in the mitochondria. Energy stored in the ATP can be carried to other parts of the organism. When ATP is broken down to ADP and phosphate, this energy is released for use in processes like active transport and muscle contraction.

ATP ⟶ ADP + phosphate + energy → biological work; ← oxidation of glucose

Breaking of the second phosphate grouping also releases energy. The remaining molecule is **adenosine monophosphate (AMP).**

1. **Using the paper models construct an AMP molecule** by matching the tabs on the adenine and ribose units at C1 and the C5 tabs on the ribose and phosphate group 1.
2. **Form ADP by bonding phosphate group 2 to phosphate 1.** A molecule of water is formed in this reaction, and energy is stored in the ADP. The wavy bond line connecting the two phosphate groups is often referred to as a **high-energy bond** because of the stored energy.
3. **Similarly, form ATP.** Again note the high energy bond between the last two phosphates.

Disassemble your molecular models and replace the components in the box. It is suggested that you keep the paper models for later study use. Complete your study by answering the following questions.

QUESTIONS:

1. List four different examples of hydrolysis considered in this exercise.

a.____________________________

b.____________________________

c.____________________________

d.____________________________

2. List two different examples of the correlation of molecular structure with function.

a.____________________________

b.____________________________

3. The oxidation of glucose to carbon dioxide and water releases energy. What is the non-living source of energy for the reduction (synthesis) of carbon dioxide by water into glucose?________

4. What molecule acts as the immediate energy source for most biological syntheses, as well as an intracellular energy carrier?____________________

5. Why is an adequate supply of the mineral phosphorus important in your diet?

6. Give the chemical formulas for the following biologically important groups.

methyl

hydroxyl

amino

carboxyl

7. Define:

high energy bond —

peptide bond —

oxidation —

reduction —

polymer —

monomer —

hydrolysis —

digestion —

synthesis —

8. The hydroxyl group could be found in carbohydrates, ______________________________,

______________________ and ____.

9. Amino groups would be found in ______________________________________ and

____________________________.

10. Carboxyl groups are found in ______________________.

CHEMICAL NATURE OF LIVING ORGANISMS

Across

2. atomic particle with a neutral charge
4. ___ molecules are composed of glycerol and fatty acids. (backward)
7. A hexose in starch is gl___.
9. Chemical ___ds occur when electrons are shared.
11. ___table of elements.
14. Carbon's ______ is six. (two words)
16. represents a bond in the models used
17. electron's charge (first three letters)
18. an electrically charged atom (backward)
19. hydroxyl group symbols
20. The mass of an elect______ is negligible.
22. are composed of three different particles
24. combines with carbon and hydrogen in carbohydrates
26. Plant cell walls are made of ___ose.
29. major component of cell membranes (backward)
30. Starch is___sized from glucose.
32. Atoms which are covalently bonded___electrons.
34. Glucose may occur in a straight chain or a___.
35. ATP = ___ + ribose + 3 phosphates
38. Starch is a___.
40. the amino acid ___ine
41. ATP loses a phosphate to become___.
43. electron gain
45. polymers of amino acids
46. Monomer units of nucleic acids are nucl___

Down

1. Carbon forms a ___ bond with oxygen.
3. 106___
5. Methane was oxidized to form ___acid.
6. Atoms of a single substance containing differing numbers of neutrons are___topes.
8. stored in ATP
10. Glycine and alanine are am___acids.
11. has a mass of 1 and a positive charge
12. sucrose
13. COOH or the___oxyl group
15. Methane with an oxygen added forms ______l alcohol.
19. OH or the hydr___group
20. 3 fatty acids combine with (first 6 letters backward)
21. ___ine, an amino acid
22. the energy storage molecule
23. electron loss or___tion
25. ATP phosphates are joined by h___ energy bonds.
27. Splitting of a molecule and adding water is hydro___.
28. ___ose, the sugar in ADP and ATP
31. protons + neutrons = atomic______
32. H joins to O with a___bond.
33. a repeating group of monomers (backward)
36. DNA, a nucleic ______
37. Starch and cellulose are___accharides. (backward)
39. 3 phosp___atoms are in ATP.
42. single units or___omers
44. Amino acids are joined together by pept___bonds. (backward)

See page 208 for answers.

CHEMICAL NATURE OF LIVING ORGANISMS

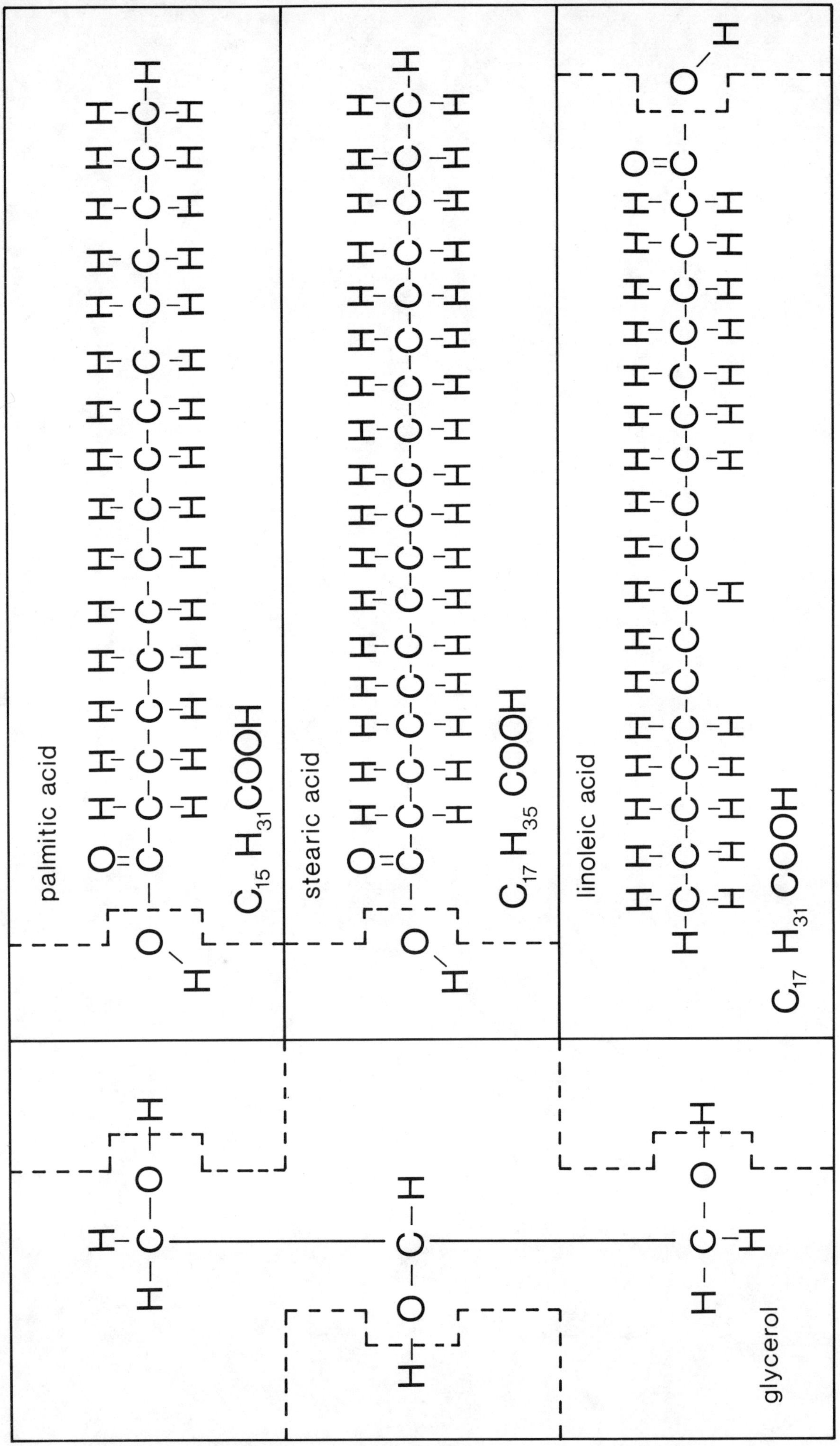

glycerol

palmitic acid

$C_{15}H_{31}COOH$

stearic acid

$C_{17}H_{35}COOH$

linoleic acid

$C_{17}H_{31}COOH$

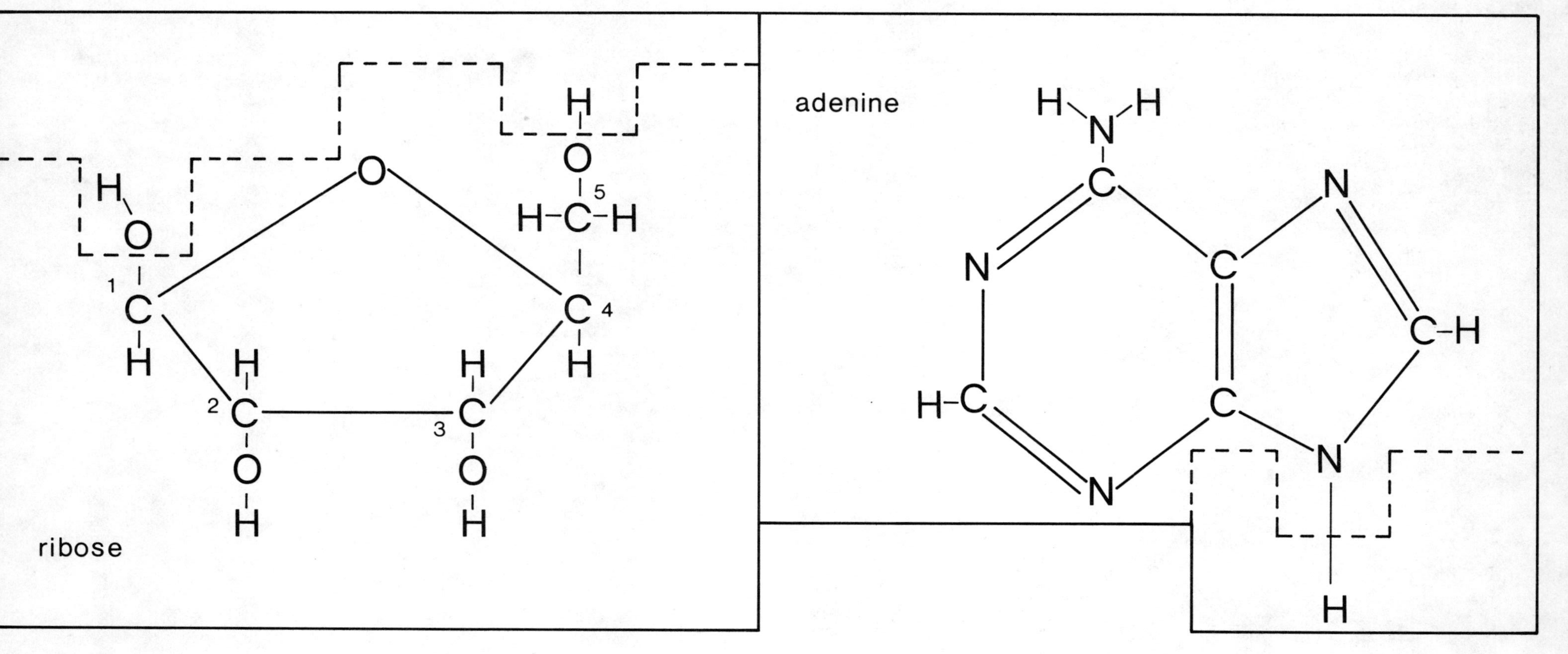
ribose
adenine

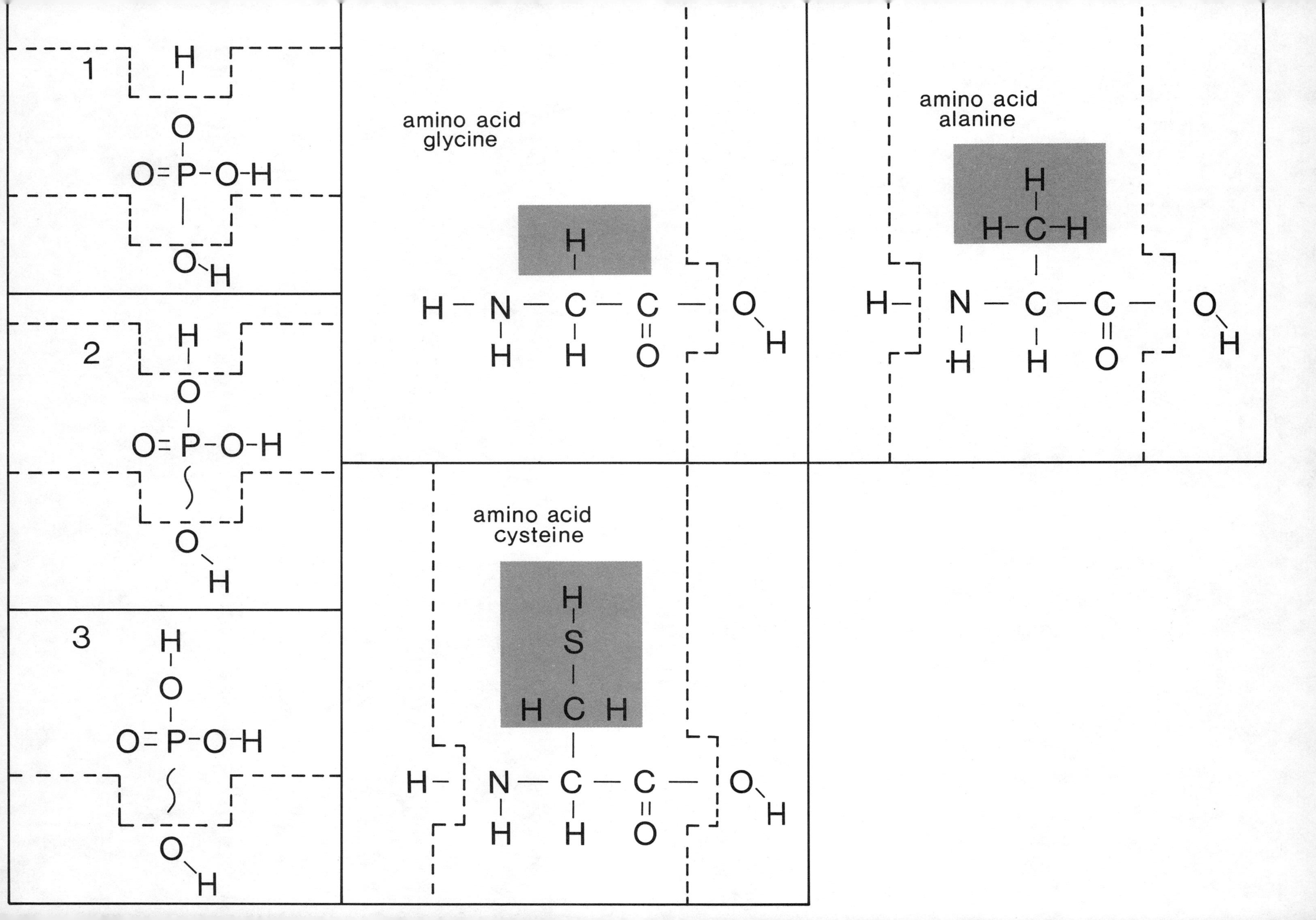
1
2
3
amino acid
glycine
amino acid
alanine
amino acid
cysteine

LABORATORY 3
ENZYMES — BIOLOGICAL CATALYSTS

OBJECTIVES: 1. To study the nature of enzymes
2. To study factors which affect enzyme reactions with the substrate

Most chemical reactions occurring within living organisms are catalyzed or facilitated by special protein molecules called **enzymes.** With a few exceptions each chemical reaction in a cell requires its own special enzyme. This rather specific relationship between an enzyme and its **substrate** (the substance upon which the enzyme acts) is described as functioning somewhat like a lock and key. Thus the enzyme must have a shape and chemical relationship complementary to that of the substrate. Some enzymes split a substrate into two parts (which may or may not be identical) as in the degradation reactions of digestion **(Figure 3.1A).** Other enzymes combine two molecules into one product in a synthesis reaction **(Figure 3.1B).** When enzyme and substrate combine, less energy is required for the substrate molecule to undergo change. When the change is complete, enzyme and product(s) separate, and the enzyme may react with more substrate molecules. Changing various conditions influences both the readiness with which the two molecules combine and the time required for transformation of substrate into product(s).

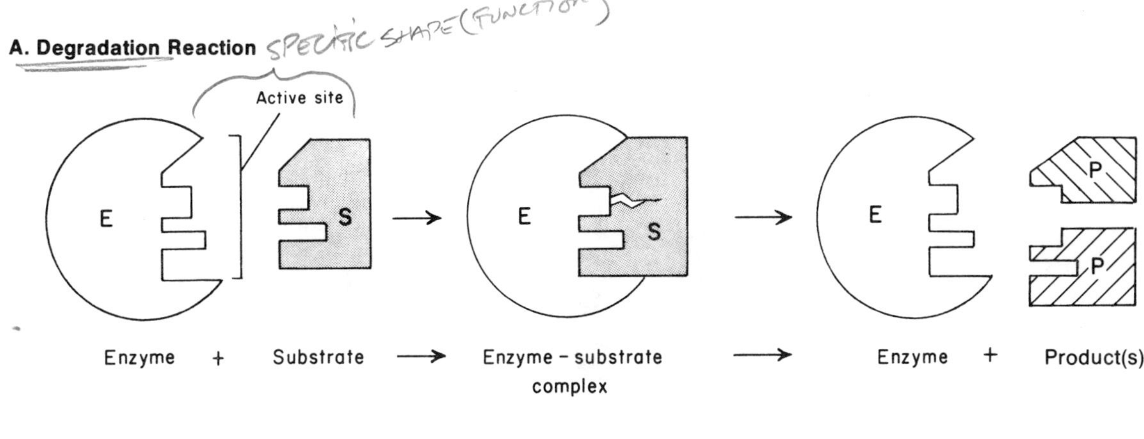

B. Synthesis Reaction

Active site
E
S
S
E
S
S
E
P

Enzyme + Substrate → Enzyme - substrate complex → Enzyme + Product

Figure 3.1. Enzyme participation in (A) degradation or digestion reaction and (B) synthesis reaction.

PRODUCTION OF ENZYMES BY LIVING ORGANISMS (Demonstration)

Petri dishes containing a starch medium were inoculated 24 hrs previously with two different types of microorganisms *(Bacillus subtilis* and *Serratia marcescens).* The two streaks on the plate's surface are the result of large numbers of bacteria actively dividing. Several drops of iodine will be placed on one end of each streak **(Figure 3.2).**

1. Observe the region where iodine was added. Remember that iodine reacts with starch to form a blue color. Thus, if the starch medium adjacent to the streak is blue, no breakdown of starch has occurred. The bacteria did not produce an enzyme capable of catalyzing the breakdown of starch.

2. If the medium adjacent to the bacterial growth does not become blue, the starch has been hydrolyzed by bacterial enzymes. Hydrolysis of starch by the enzyme would lead to the production of ________.

3. Results:

Organism	**Starch Hydrolysis**
B. subtilis	________
S. marcescens	________

4. Which organism could use starch as a food or nutrient source?________________

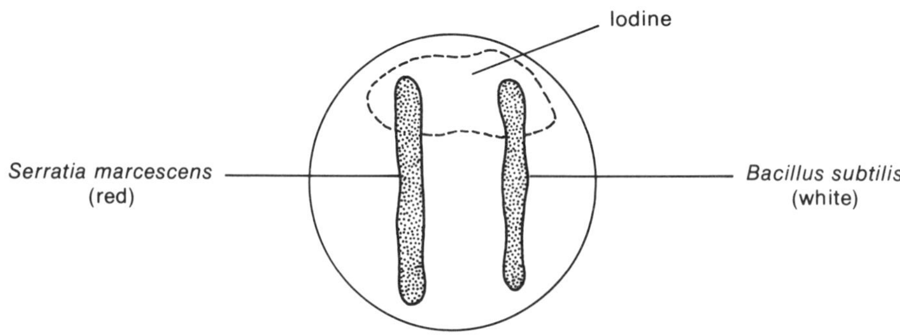

Figure 3.2. Petri plate showing testing of two organisms for the ability to hydrolyze starch.

PROTEIN NATURE OF ENZYMES (Demonstration)

The enzyme produced by the bacteria in the previous demonstration was a protein. Proteins contain numerous peptide bonds linking the various amino acids into a chain. Sections of this amino acid chain may be connected with each other by additional relatively weak bonds, or two different chains of amino acids may be linked together by these bonds. This produces a folded, complex molecule, rather than a long chain. If these bonds are broken, the molecule unfolds **(denatures)** and the enzyme will not function. Breakage of these bonds may occur at elevated temperatures, with pH changes, with certain chemicals, etc.

A Biuret test will be performed to demonstrate the protein nature of the enzyme you will use today. The procedure will be the same as previously used in the Scientific Method exercise.

Tube	Final Color	Interpretation
Control (1 ml H_2O)	BLUE	
Experimental (1 ml 0.6% diastase)	CLEAR	

END PRODUCT OF THE REACTION

In the remainder of the exercise, effects of various factors on enzyme activity will be determined, using diastase (a starch-splitting enzyme). The substract acted upon will be starch. Our overall equation could be written as:

substrate + enzyme ⟶ substrate-enzyme complex ⟶ product + enzyme

or, more specifically,

starch + diastase ⟶ starch-diastase complex ⟶ maltose + glucose + diastase.

Would this be an example of a synthesis or a degradation reaction? ____________

To indicate progress of the reaction, a dilute iodine solution will be used. Starch plus iodine results in a blue color. Maltose and glucose plus iodine do not become blue. Thus the presence of the blue can be used as an indicator of reaction progress.

Students will work in groups of four. **CLEAN GLASSWARE IS ESSENTIAL,** as even traces of diastase left in a tube or pipette may ruin your test results. Thoroughly rinse all tubes, pipettes and syringes before using them to transfer another solution.

The following procedure will be used throughout today's experiment: (1) label one pipette or syringe for use with diastase **only;** (2) always add solutions in the order shown in the tables; (3) add diastase last, immediately shake tubes to mix and **begin timing at once.**

1. Prepare two tubes as follows.

Tube	Starch	Iodine	Water	0.4% Diastase
A	3 ml	1-2 drops	1 ml	0
B	3 ml	1-2 drops	0	1 ml

2. Remember to begin timing immediately after addition of the enzyme to the experimental tube.
3. Record time required for tube B to become completely colorless. (It will first become pinkish and then completely clear.) An additional drop of iodine may be added to be certain reaction is complete. If not, continue until it clears again.
4. **Save both tubes for part 6.**
5. Loss of blue color indicates starch has been hydrolyzed to another compound by the enzyme. Since starch is composed of repeating units of glucose attached to each other, breakdown of the starch might be expected to produce glucose (or maltose, which is two glucose molecules hooked together). To determine if this did occur in tube B, you will perform a test for the presence of glucose and maltose.
6. Prepare a known positive by placing 3 ml of glucose solution in a tube (labeled C). (Be certain to rinse pipette or syringe thoroughly before using it in other solutions.) Add 3 ml of Benedict's solution to this tube and to tubes A and B (from Section 1 above.)
7. Place all 3 tubes in a boiling water bath for 8 minutes. Remove and let cool 3 minutes.
8. Observe for the presence of a precipitate. In the Benedict's test, the blue cupric ion (Cu^{++}) is converted to the red cuprous ion (Cu^{+}) by the aldehyde group of the sugar. The Cu_2O which results is a red precipitate. The presence of varying quantities of reacting sugar may be shown by precipitates varying in color from blue-green to red according to the following.

Blue: negative	Yellow: 2+
Blue-green: trace	Orange: 3+
Green: 1+	Red: 4+

9. Results:

Tube	Contents	Time Required to Become Colorless	Reaction with Benedict's
A	starch, H_2O, Benedict's	--------------	blue (no sugar)
B	starch, diastase Benedict's		aqua green
C	glucose, Benedict's	--------------	V-8 Red (no starch)

EFFECT OF INHIBITORS ON ENZYME ACTIVITY

Enzyme activity may be blocked by certain chemicals. These chemicals may combine either with the enzyme or with the substrate, preventing enzyme and substrate from coming together in the lock and key manner.

1. To a tube labeled inhibitor, add 1 ml of 2% copper sulfate and 1 ml of 0.4% diastase.
2. Swirl the contents to mix and allow to stand at least 15 minutes. Exact timing is not essential here, as what you are doing is allowing sufficient time for the copper sulfate to combine with the enzyme.
3. Add 3 ml of starch solution and 4 drops of iodine. Mix and **note time.**
4. Observe periodically for color change. If not colorless after 30 minutes, discontinue readings and record as > 30 minutes. You may continue with the next section of the exercise and check on this tube at intervals.
5. Results:

EFFECT OF TEMPERATURE ON ENZYME ACTIVITY

1. Prepare the following.

Tube	Starch	Iodine	Incubation Temperature
I	3 ml	1-2 drops	4° C (Ice Bath)
II	3 ml	1-2 drops	22° C (Room Temp.)

2. Leave at designated temperatures for ten minutes to equilibrate.
3. Leave the tubes at the designated temperature and add 1 ml of 0.4% diastase to each. Swirl and **immediately note time.** Observe for color change at intervals. If no change is observed after 30 minutes, discontinue readings and record as > 30 minutes.

4. Results:

Tube	Reaction Temperature	Time Required to Become Colorless
I	4° C	______
II	22° C	______

Plot time required for enzyme action at the two temperatures.

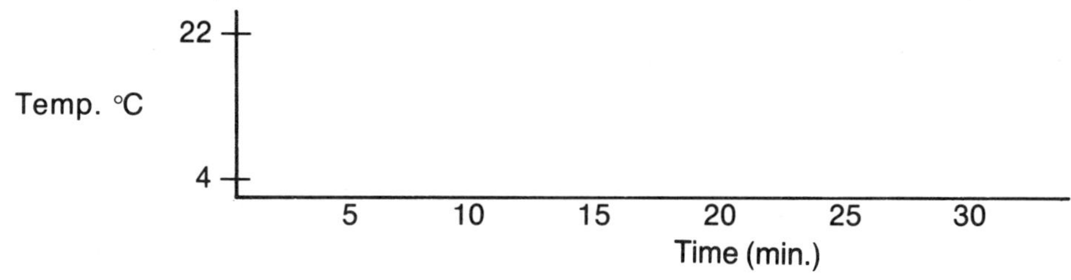

Denature - break apart, become useless

EFFECT OF CONCENTRATION ON ENZYME ACTIVITY

1. Prepare tubes as shown.

Tube	Starch	Iodine	Diastase 0.2%	Diastase 0.6%
I	3 ml	1-2 drops	1 ml	0
II	3 ml	1-2 drops	0	1 ml

2. The same pipette may be used for both diastase solutions, providing the weaker solution is pipetted first.
3. Swirl contents, **note time** and observe for color change.
4. Results:

Tube	Diastase Concentration	Time Required to Become Colorless
I	0.2%	______
II	0.6%	______

Plot times on chart below. (Add a point for the time determined previously for the 0.4% concentration in number 4 under effect of temperature).

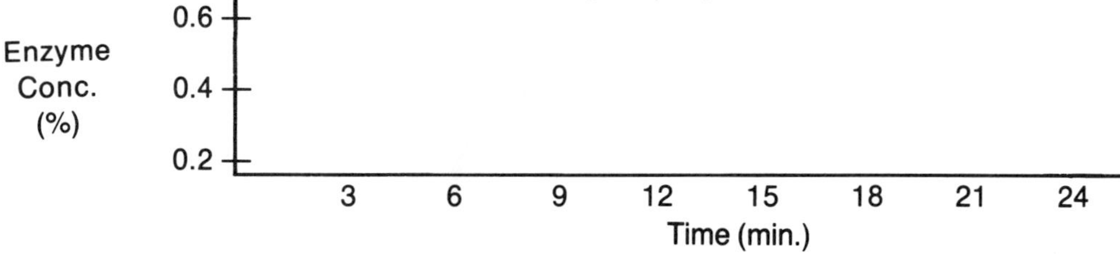

CONCLUSIONS AND QUESTIONS

1. Enzymes (are, are not) produced by living organisms. All living organisms (do, do not) produce the same enzymes.

2. Diastase reacts with starch to produce ______________

(as determined by tube ________).

3. Diastase (is, is not) a protein.

4. How do inhibitors prevent an enzyme from functioning?

5. The lower the temperature, the ________ the time required for completion of the reaction. If the temperature were raised very much above 37°C, what results would you expect? Explain.

6. The greater the concentration of the enzyme, the ________ the time required for completion of the reaction.

7. On the basis of the chart you plotted, what time would you expect for a reaction using 0.1% diastase? ______________

8. On the basis of your data, what would you anticipate as the reaction time at 15° C? ________ at 35° C? ________

LABORATORY 4

MICROSCOPE USE AND PHYSICAL PROCESSES IMPORTANT TO LIFE

OBJECTIVES: 1. To learn proper care and use of the compound microscope.
2. To investigate some physical processes important in living organisms.

The compound microscope is a system of lenses and mechanical parts arranged to allow minute movements of the lenses in relation to the specimen. It is a precision instrument and only proper care and usage will maintain it in a condition in which it can be of benefit to the laboratory investigator. In this exercise you will become familiar with the various parts of a microscope and their functions, as well as how to use and provide "tender loving care" for this instrument.

Carefully remove the assigned microscope from its cabinet and take it to your desk, carrying it in the manner shown by your instructor. Examine the microscope and compare it with **Figure 4.1.** Study the parts carefully and relate them to their functions.

MICROSCOPE PARTS AND FUNCTIONS

1. **mirror.** A flat mirror will change the direction of a light beam. A concave mirror does this, but also focuses the beam to provide greater intensity of illumination.

2. **objectives.** Objectives contain several lenses. Observe the numbers inscribed on the sides of the objectives. One number gives the approximate magnifying power of the lens (10X or 43X) and the other gives the focal length (16mm or 40mm). The 10X means that the magnifying power is on the order of ten diameters (or the specimen is enlarged approximately ten times). Objectives are mounted on a rotating nosepiece, held in position by a spring clip so that moving it into position produces an audible "click."

3. **eyepiece** or **ocular.** The eyepiece contains the ocular lens and further enlarges the image formed by the objective. Total magnification of the microscope is the product of the magnifying power of the objective and ocular. Thus, with a 10X ocular and a 10X objective, the total magnification would be 100X. Rotation of the eyepiece and its contained pointer allows the pointer to be placed on a particular part of a specimen that you want to show to the instructor or to a fellow student in order to discuss the object in question.

4. **aperture disc or iris diaphragm.** Located immediately below the stage on some microscopes is a wheel or disc containing a graduated series of holes. These are utilized to regulate the contrast between the specimen and its surroundings. Some microscopes have an iris diaphragm to control contrast. Moving a lever will regulate the size of this opening, much like the diaphragm on a camera or the iris of your eye. In addition, in many scopes there is a condenser lens located in this region which also influences light reaching the eye.

5. **coarse adjustment knob.** This knob is connected to a rack and pinion gear on which the body tube of the microscope moves up and down. It is used whenever larger movements of the body tube are required.

6. **fine adjustment knob.** This knob raises or lowers the body tube through a limited distance, giving very precise control over lens positions. It is used to sharpen focus.

7. **stage.** The stage is the platform upon which the slide is placed. Slides are held in position by stage clips. Some stages are equipped with mechanical holders which allow more precise movements of the slide.

8. **base.** The base is the horseshoe-shaped portion upon which the instrument rests.

9. **arm.** This is the part of the instrument which supports the adjustment knobs and the **body tube.**

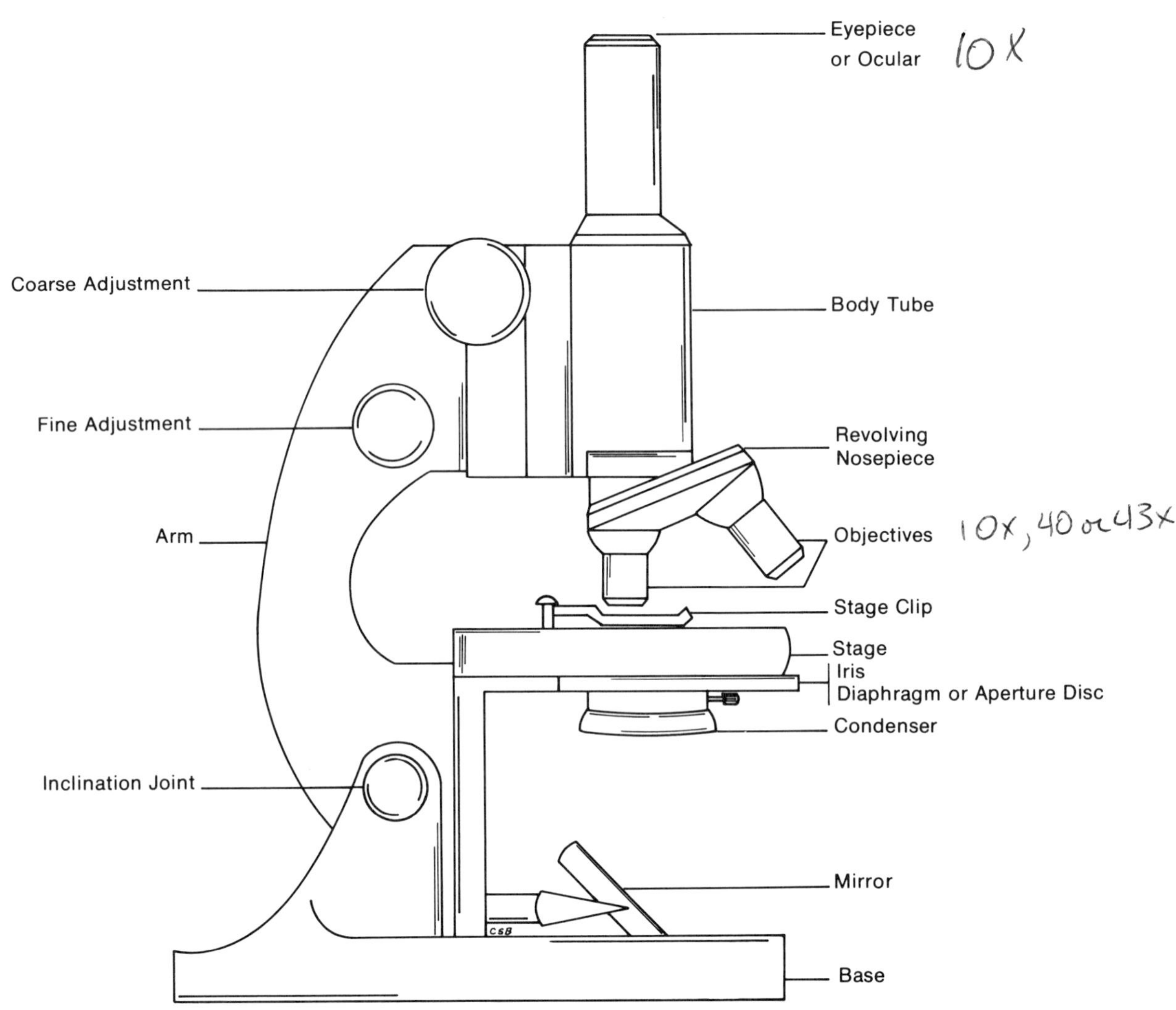

Figure 4.1. The microscope.

MICROSCOPE CARE

ALWAYS	REASON
1. Carry the scope **by its arm** in an **upright position** with one hand under the base.	1. Prevents eyepiece falling out or jamming scope against edge of table.
2. Clean eyepiece, objectives and mirror **with lens paper** before and after use.	2. Prevents scratching of lenses.
3. **Look from the side while lowering objective** toward slide.	3. Prevents striking of slide with objective and possible loosening or breaking of lens.
4. Before storing check to see that: body tube is upright, 10X objective is in place, body tube is lowered, slide is removed, stage cleaned.	4. Prevents possible damage to lenses (low power objective lock prevents striking of stage with objective). Proper storage prevents temper tantrum by instructor.
5. Report any damage immediately.	5. Prevents instructor from "chewing out" the wrong person — you.

MICROSCOPE PROCEDURE

Setting up the low power objective (10X).

1. If desired, body tube may be tilted **slightly.** Avoid excess inclination as this may lead to the microscope being knocked over. Do not incline if using wet preparations (or the specimen may end up in your lap!).

2. Place the low power (10X) objective into position. Look through the eyepiece and adjust the concave mirror to obtain maximum illumination. **Always start with this objective.**

3. Place the colored threads slide on the stage and secure it with stage clips. (Do not allow clips to rest on the coverslip.)

4. Hold your head to the side of the microscope and while watching the objective, lower the body tube with the coarse adjustment knob until the lens is just above the slide. Usually a lock will prevent the low power from going too far, unless the slide is thick. **Never assume that this lock will protect you — always look.**

5. Looking through the eyepiece, raise the tube with the coarse adjustment until the specimen comes into focus.

6. Select proper aperture by rotating aperture disc (or adjusting diaphragm) until the image is as brightly illuminated as is comfortable for the eye, and object edges are sharp and distinct. Some adjustment of the mirror may also be necessary. If your scope has a condenser, your instructor will explain how it should be adjusted.

7. When the specimen is centered, in sharp focus and properly illuminated, go to the next step.

Setting up the high power objective (43X).

1. **Always locate specimen and focus with low power objective before attempting to obtain greater magnification.** If you can't find the specimen on low power, going on to the next objective won't help!

2. Center the part of the specimen you wish to examine. The high power objective will cover less area on the slide and failure to center the specimen may mean that it will not be in the field of view with the high power objective.

3. Use fine adjustment only.

4. The high power objective is to be used to examine very thin preparations. Never attempt to use it on thick whole mounts, since a broken slide and/or damaged objective will be the usual result.

5. Most of the microscopes are **parfocal** (the image will be in focus with both objectives with only minor adjustments). **To change objectives, observe the objective from the side as you slowly rotate the nosepiece until the desired objective clicks into place.** (If unable to do this without the objective striking the slide, the microscope is not parfocal — proceed with step 6.) It will probably be necessary to sharpen the focus with the fine adjustment and to increase the amount of light.

6. **Procedure if microscope not parfocal**

   a. Raise the body tube slightly and rotate the high power objective into place. Looking at the objective from the side (NOT through the eyepiece), lower the tube with the coarse adjustment until the objective almost touches the coverslip.

   b. While looking through the microscope, slowly raise the tube with the coarse adjustment until the object comes into focus. Sharpen the focus with the fine adjustment and adjust the light.

Things that can go wrong.

1. The fine adjustment may jam before the object comes into focus. Correct this by raising the tube clear of the slide and turn the fine adjustment knob 4 to 6 revolutions the other way. Repeat procedure for focusing.
2. A dimly lighted indistinct image may result from any of the following causes:
   a. Aperture too small or condenser too low.
   b. Using the flat side of the mirror.
   c. Dirty eyepiece.
   d. Dirty objective.
   e. Trying to examine a thick object or a mass of material under too much magnification.

USING THE MICROSCOPE TO OBTAIN INFORMATION

Depth of Focus. As the thickness of specimen in focus at any one time is quite small, it is necessary to change focus to obtain the relative position of structures in the third dimension (along the optical axis). Place the colored thread slide on the stage, and with low power find a place where all three colors of thread are visible. Focus independently on each color (using the fine adjustment). Determine the order of the threads from top to bottom and record the data. (Your instructor will confirm your results.) This type of information is essential in certain laboratory procedures, such as identifying malarial parasites inside of red blood cells.

Color of thread on top __red__

Color of thread on bottom __yellow__

Field Size and Measurements. If the field size produced by the objective-ocular combination of a microscope is known, it is possible to estimate the size of relatively large specimens by determining how much of the field of view is occupied by the specimen. The diameter of the low power field is approximately 1.4mm; the diameter of the high power field is approximately 0.36mm. Estimate the diameter of the red thread on your slide. __1.4 +__ mm.

If a "critter" is 0.04mm long, how much of the low power field would it occupy? ________ the high power field? ________

Magnification. Magnification of eyepiece __10x__, low power objective __10x__, high power objective __40__. Total magnification with low power __100x__; total magnification with high power __400__.

In the following table, place a checkmark under increase or decrease to indicate the effect that occurs when the total magnification is increased from 100X to 430X. If necessary, actually observe each using your microscope and the letter "e" slide.

	Increase	*Decrease*
Field of view	______	______
Light intensity	______	______
Depth of focus	______	______

Specimen orientation. Set up the letter "e" slide on **low power.** Examine the letter on the slide and compare its orientation on the slide as it appears to the unaided eye with its orientation when observed through the eyepiece.

Letter as seen through the microscope ______

Letter as seen with unaided eye ______

If the slide is moved to the right on the stage, which way does it appear to move when viewed through the microscope? ______

PHYSICAL PROCESSES IMPORTANT TO LIFE

Certain physical processes are biologically important because they partially account for the movement of water, gases and dissolved substances into and out of living cells (through membranes) and also within the cytoplasm.

Brownian movement. Very small particles suspended in a fluid can be seen to dance about in a random manner. The erratic motion of the particles is caused by the impact of molecules of the solvent striking the suspended particles. The heat energy of the solvent molecules manifests itself as their random motion.

1. Shake the carmine solution and place one drop on the center of a clean glass slide. Hold the coverslip at a slight angle and touch the lower side to the drop of carmine. Slowly lower the high edge, taking care not to trap air bubbles beneath the coverslip. A dissecting needle makes a convenient tool to support the coverslip as you lower it. (It's smaller and it doesn't leave greasy fingerprints!) This preparation is known as a **wet mount.**

2. **With the microscope stage level,** place the slide on it and observe the drop under both low and high power. It may be necessary to use less light to best observe the Brownian motion of the dye particles. Do you notice any difference in the motion of particles of different size? Could this process serve as a means of locomotion for a small fish? ______

Why or why not? ______

Diffusion. The movement of substances from regions of higher concentration to regions of lower concentration as a result of the heat (kinetic) energy of the molecules is termed **diffusion.** This physical process is involved in the movement of some substances into and out of cells.

1. **Diffusion of a gas in a gas (demonstration).** The apparatus will be set up as shown in **Figure 4.2.** One cotton plug, dampened with hydrochloric acid (HCl), and another, dampened with ammonium hydroxide (NH_4OH), will be inserted simultaneously into opposite ends of the glass tube. As the liquids evaporate, the two gases diffuse through the tube until they meet. A white ring will appear at the point where the two chemicals react with each other. Draw the location of this ring indicating contact of the two gases.

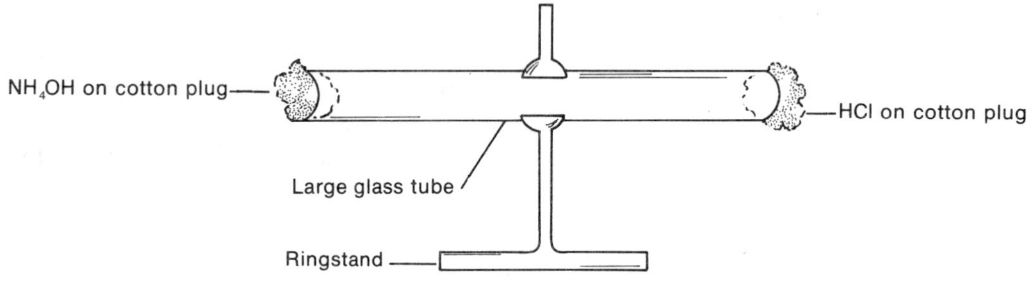

Figure 4.2. Gas diffusion apparatus.

What caused the molecules of HCl and NH_4OH to diffuse through the air in the glass tube?

What served as the solvent?

air

Why didn't the ring form halfway between the two cotton plugs?

Atomic Weight (as related to speed)

2. **Diffusion of a solute through a gel** (demonstration). A cork borer was used to cut four holes in the agar in a petri dish **(Figure 4.3).** Each hole will be filled with a different solution: hydrochloric acid, methylene blue, potassium ferricyanide and silver nitrate. Label the drawing to indicate the solution in each numbered well. After the solutions are added, allow the plate to stand undisturbed for 20 to 30 minutes. Sketch the results on the figure and record the distance in millimeters each of the substances migrated through the agar.
NOTE — HCl is colorless. Hold the petri dish at various angles to the light in order to see the ring formed by the migration of the chloride ion.

Substance	Migrating Ion	Relative Weight	Distance Migrated (mm)
hydrochloric acid	Cl^-	35	
potassium ferricyanide	$Fe(CN)_6^{\equiv}$	212	
silver nitrate	NO_3^-	112	
methylene blue	aniline	>400	

From the results recorded, what can you conclude about the relationship of diffusion rate of a molecule to its molecular weight?

3. **Diffusion of a solid in water.** Fill one of the flasks (or beakers) on your table with hot tap water and label it. Fill the other container from the cold water tap and label accordingly. Allow both to sit undisturbed on your table until the water currents subside (at least two minutes). Put a **few** crystals of potassium permanganate ($KMnO_4$) into each. Attempt to put equal quantities of crystals into each container and do not disturb the containers for the remainder of the laboratory period. Periodically observe both containers and compare the rates of diffusion.

What causes the purple color of the $KMnO_4$ to spread throughout the water even though it is not disturbed?____________________

Was the diffusion rate of $KMnO_4$ in warm water different from that in cold water?__________What is the relationship between temperature and diffusion rate?____________________

What actually caused the difference in diffusion rate?

____

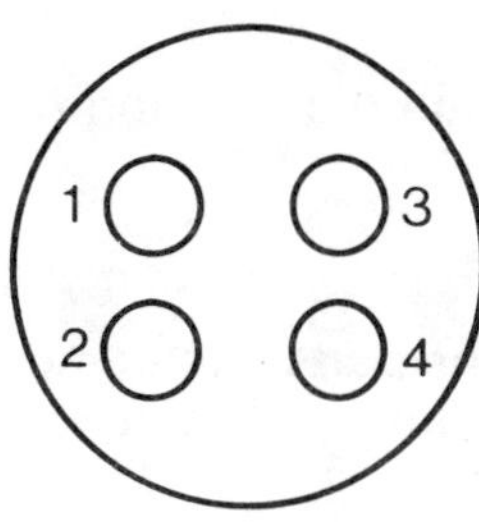

Figure 4.3. Plate prepared to show diffusion of a solute through a gel.

Osmosis. The movement of water through a selectively permeable membrane in response to concentration differences on the two sides of this membrane is termed **osmosis.** Like other substances, water moves (diffuses) from regions of higher concentration to regions of lower concentration. The cell membrane is selectively permeable; that is, it permits the passage of some molecules, such as water, while preventing the passage of others. The concentration of water in a solution is determined by the concentrations of the various substances dissolved in it. Thus the concentration of water in a strong salt solution is less than the concentration of water in a weak salt solution. Whether water will move through the cell membrane into or out of the cell depends, then, on the difference in the water concentration within the cell and in the fluid which bathes it.

1. **Definitions**

Isosmotic solutions contain the same concentration of solute (dissolved particles) as do the cells.
Hyperosmotic solutions contain a higher concentration of solutes than do the cells.
Hypoosmotic solutions contain a lower concentration of solutes than do the cells.

2. **Osmosis in non-living systems** (demonstration). At the beginning of the laboratory period two osmometers or thistle tubes with selectively permeable membranes closing one end will be set up. One tube is filled with pure molasses, while the other contains a 50% molasses-water solution. The solutions extend for a short distance up into the tubing. The membrane-covered end of each is immersed in a beaker of water. The height of the two sugar solutions is marked at the beginning of the period and each column should be measured to the nearest millimeter at least three times during the lab period at either 20- or 30-minute intervals.

Tube with Pure Molasses	**Tube with 50% Molasses — 50% Water**
_____minutes -______mm	______minutes -______mm
_____ minutes -______mm	______ minutes -______mm
_____ minutes -______mm	______minutes -______mm

Why does the fluid rise in the glass tubes?

How could you reverse the process so that the solution would descend in the tube without changing the thistle tube and its contents?

Explain any differences in the rate at which the solution rises in the two tubes.

Sketch the osmotic apparatus and label the parts, including labels for the contents of beakers and tubes. Indicate the height of the sugar solution at the beginning and end of the observations.

3. **Osmosis in living systems.** Refer to Laboratory 1 and the changes that you observed in the potato core after immersion in a concentrated salt solution. The core lost water because it was immersed in a (hypoosmotic, hyperosmotic) solution. The potato cells were (hypoosmotic, hyperosmotic) to the salt solution.

Another living system which will be used to show the effects of variation in osmotic concentration on living cells is the *Elodea* leaf. Remove the tip of an *Elodea* leaf, place it in a drop of water on a slide, and cover with a clean coverslip. Upon examining the leaf with both low and high power, you will note that these cells contain numerous green bodies called chloroplasts distributed more or less uniformly throughout the cell.

Raise the coverslip, blot the water with absorbent paper, and add a drop of hyperosmotic (20% salt) solution. After several minutes, observe the cells. Those cells nearer the margin of the leaf should react first. Describe the behavior of the cell's contents.

Remove the coverslip, rinse the leaf with water, blot and remount the leaf in hypoosmotic (distilled water) solution. Again observe the cells for changes after several minutes. Describe these changes.

Sketch three cells; illustrate with arrows the direction of water movement for each solution type used (iso-, hypo- and hyperosmotic).

What would happen to your cells if you slept in a vat of sea water? Explain.

LABORATORY 5

THE CELL AS A UNIT OF STRUCTURE

OBJECTIVES: 1. To observe basic similarities of structure found in all cells.
2. To observe differences between plant, animal and moneran cells.

The basic unit of life is the cell. A cell consists of a number of structural parts, many of which are visible with a light microscope, and some of which are so small that they can be observed only with the electron microscope. Multicellular organisms contain many different cell types, each type specialized for its specific function. Cells range in size from <1 micron in diameter, the size of some bacteria, to 77 mm in some bird eggs. Most cells, however, have a diameter of 5 to 15 microns.

Cell boundaries and internal organization may be used as a basis for categorizing cells. **Eucaryotic cells** have nuclear material enclosed by a nuclear membrane and characteristically contain a number of membrane bounded organelles (or "little organs"). With the exception of the plant's photosynthetic structures (chloroplasts), organelles are essentially the same in plant and animal cells. In addition, plant cells have cell walls, generally composed of cellulose. The outer boundary of an animal cell is the cell membrane. Members of the kingdom Monera (bacteria and blue-green algae) have cell walls also, but they differ in chemical composition from the plant cell wall. The monerans are termed **procaryotic cells** because the hereditary material of the cell is not separated from the rest of the cell by a membrane. Procaryotic cells also lack most of the organelles associated with eucaryotic cells.

CELL STRUCTURE

In **Figure 5.1** are diagrams of a plant and an animal cell showing structure and organelles. Study the models in the lab, the following descriptions, and your text and label the indicated parts of the diagrams.

1. The **nucleus** is a membrane-bounded structure which functions in regulation of cellular activities, including division. A number of internal features may be distinguished.
   a. The **nuclear envelope** forms the outer boundary. With an electron microscope it is seen to consist of a double-layered membrane, containing pores.
   b. The **chromosomes** are the hereditary material of the cell. They are seen as individual structures in dividing cells, but in non-dividing cells appear as scattered granular material called **chromatin.**
   c. **Nucleoli** are dense spherical structures involved in the synthesis and assembly of ribosomes.
   d. The fluid interior of the nucleus in which other parts are suspended is the **nucleoplasm.**
2. The **cytoplasmic matrix** is the apparently structureless "filling" of the cytoplasm.
3. Scattered throughout the cytoplasmic matrix is a membranous network, the **endoplasmic reticulum,** visible through the electron microscope. The endoplasmic reticulum may function as transport channels within the cell, and may also be involved in protein synthesis, since it is often studded with ribosomes.
4. **Mitochondria** are usually oval organelles in which food molecules are dismantled to provide energy for the life processes of the cell. They appear as small dots or threads with the light microscope, but electron micrographs reveal the double membrane structure.
5. **Ribosomes** appear under the electron microscope as tiny particles approximately 1 millimicron in diameter. These organelles are the sites of protein synthesis.
6. The **Golgi apparatus** appears as stacks of membranous sacs using an electron microscope. Apparently it is involved in several cell activities, including production of secretory products.

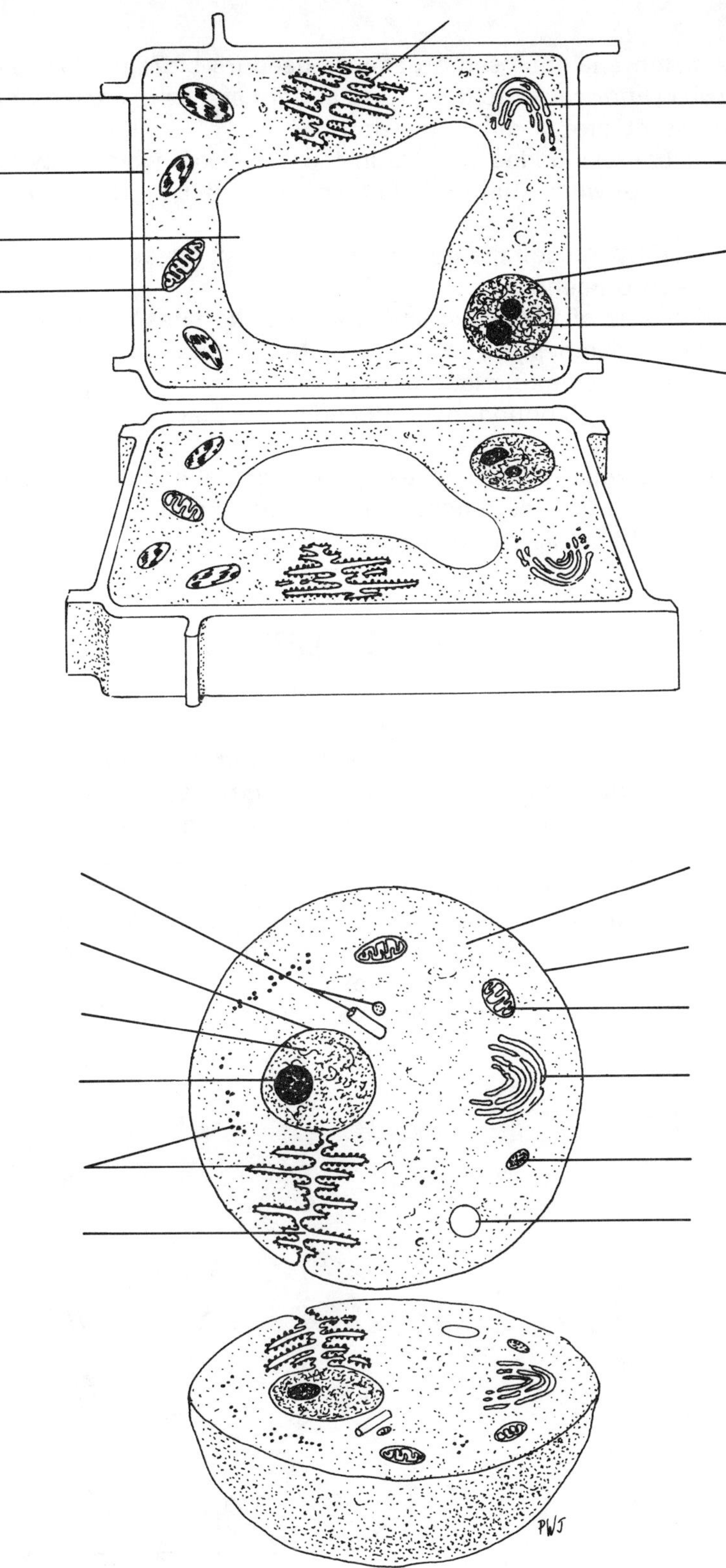

Figure 5.1. Diagrams of "typical" plant (top) and animal (bottom) cells.

7. **Lysosomes** are membrane enclosed organelles used for storage of digestive enzymes.

8. **Centrioles** are two cylindrical structures, lying at right angles to each other and usually close to the nucleus. They are associated with cell division.

9. **Vacuoles** are membrane-bounded vesicles in the cytoplasm which may surround food particles being digested by the cell, or which may contain fluid involved in regulating the water content of the cell.

10. The cell is bounded by the **cell membrane** which regulates exchange of substances between the cytoplasm and its environment.

11. **Plastids** are structures of various shapes that are the sites of photosynthesis or, in some cases, storage of materials synthesized by the cell. Those that contain chlorophyll are specifically called **chloroplasts.** They are characteristic of plants.

12. The **cell wall** is a fairly rigid boundary produced by the plant cell and lying just outside the cell membrane.

13. Many cells possess specialized organelles which enable the entire cell to move about, or which move materials over the cell surface. These hair or whip-like structures extending from the cell surface are called **cilia** and **flagella** respectively.

TYPES OF CELLS

Plant Cells Without Chloroplasts

Remove a small section of onion and, while holding it with the concave side facing you, snap it backwards. This should produce a ragged edge consisting of transparent, filmy epidermis which is one cell layer thick. With your forceps, remove a small piece of this epidermis and prepare a wet mount, using a drop of iodine solution rather than water. Avoid folding or wrinkling the epidermis.

Examine the preparation under low and high power, noting the appearance of **nucleus, nucleolus, cytoplasm, vacuole,** and **cell wall.** It may be helpful to reduce the light for an increase in contrast. Refer to **Figure 5.2.** Under high power, change focus with the fine adjustment recalling what you learned about depth of focus in the microscope exercise.

Sketch and label a representative onion cell in two dimensions as it appears through your microscope.

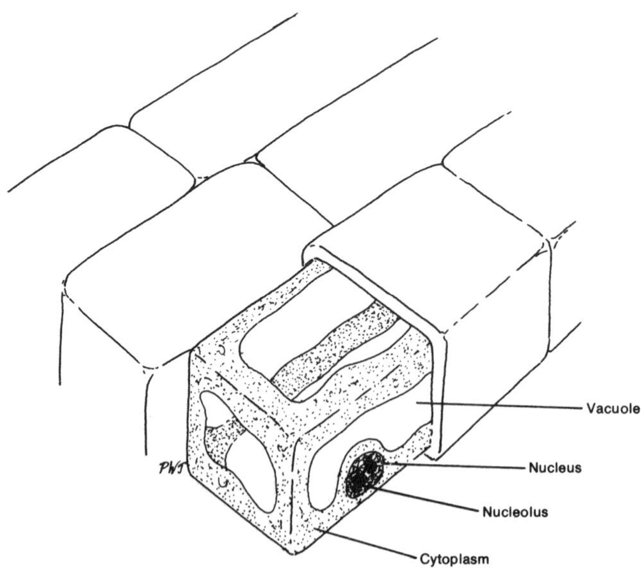

Figure 5.2. Three dimensional diagram of onion epidermal cell.

Plant Cells With Chloroplasts

Prepare a wet mount using the tip of an *Elodea* leaf. Focus on an area near the tip adjacent to the midrib of the leaf. Note the cell walls. The cell membrane is closely pressed to the wall and cannot be distinguished. The rounded, green structures located within the cytoplasm are **chloroplasts.** Observe these structures closely. Cyclic motion of the cytoplasm is known as **cyclosis,** and movement of the chloroplasts will allow you to see the effect of this cytoplasmic streaming within the cell. Nuclei are difficult to see. An area devoid of chloroplasts may be present in the cytoplasm. This is a vacuole. Draw and label a representative cell.

ANIMAL CELLS (*Amphiuma* Liver)

These slides were prepared by making very thin, uniform slices of the liver of the amphibian *Amphiuma.* The slices were then stained and permanently mounted under a coverslip. Observe the section with low and high power, noting the **nucleus, nucleoli,** and **cytoplasm.** The nucleus is typically stained blue and the cytoplasm pink. The cell membrane forms the outer boundary of each cell. Note the shape of the cells. These sections also contain **glycogen,** which is seen as reddish-brown granular material in the cytoplasm. Glycogen is a carbohydrate storage product. This polymer of glucose corresponds to starch in plants. Draw and label the parts of 1 or 2 representative cells.

BACTERIA (PROCARYOTIC CELLS)

On the prepared slide of bacterial types focus first with low power, then with high. Find representatives of the three morphological types **(Figure 5.3).** Also, look for living bacteria in the pond water sample you will study next. Some bacteria use flagella for movement. Others lack flagella and show only Brownian motion, not true motility. How can you distinguish between Brownian motion and true motility? ______________________________

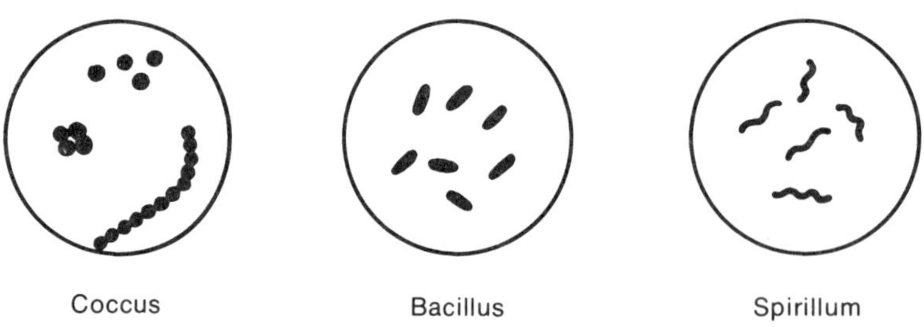

Figure 5.3. Morphological types of bacteria.

LIVE CELLS — POND WATER SAMPLE

A fascinating world of microscopic life exists within a single drop of pond water. Place a drop from the pond water culture on a clean slide, including a bit of the solid material. Cover with a clean coverslip. Since many of the organisms are nearly transparent, a vital stain, methylene blue, can be added to the water. This kind of stain will color certain parts of the organisms without killing them, enabling you to more clearly see cell structures. Place a drop of 1:10,000 methylene blue at the edge of the coverslip on the slide. Touch a piece of paper toweling to the opposite edge, causing the stain to flow under the coverslip. Continue until the liquid absorbed by the towel is faintly blue. Scan the slide under low power, observing as many different kinds of organisms as possible. Those described in the following paragraphs and illustrated in **Figure 5.4** are some of the ones commonly seen. After locating an organism, switch to high power to observe it. Over a period of 5-10 minutes, the vital stain will gradually be absorbed and you should see internal structure more clearly. How many structures and organelles can you see? Remember that some of these organisms are unicellular, while others are small, multicellular organisms. Still others may be filaments composed of cells in chains.

The protozoan *Vorticella* has a ring of cilia around the broad end of its bell-shaped body. It attaches to the substrate by a slender stalk containing a spiral contractile fiber. When undisturbed, the bell is expanded on the end of the stalk, with the cilia in rapid motion. A slight disturbance causes the stalk to contract, and the bell folds over the circlets of cilia. *Vorticella* feed chiefly on bacteria.

Rotifers can be recognized immediately by the presence of a crown of cilia, which serves as the chief organ of locomotion and also to bring food into the mouth. In some forms, the beating of the cilia gives the appearance of a revolving wheel. The tapering foot anchors the body during feeding. Rotifers are unusual since their bodies are not divided into distinct cell units, but consist of a protoplasmic mass containing a number of nuclei. Cell walls are present in embryonic stages, but later disappear. The number of nuclei present is constant within a given species.

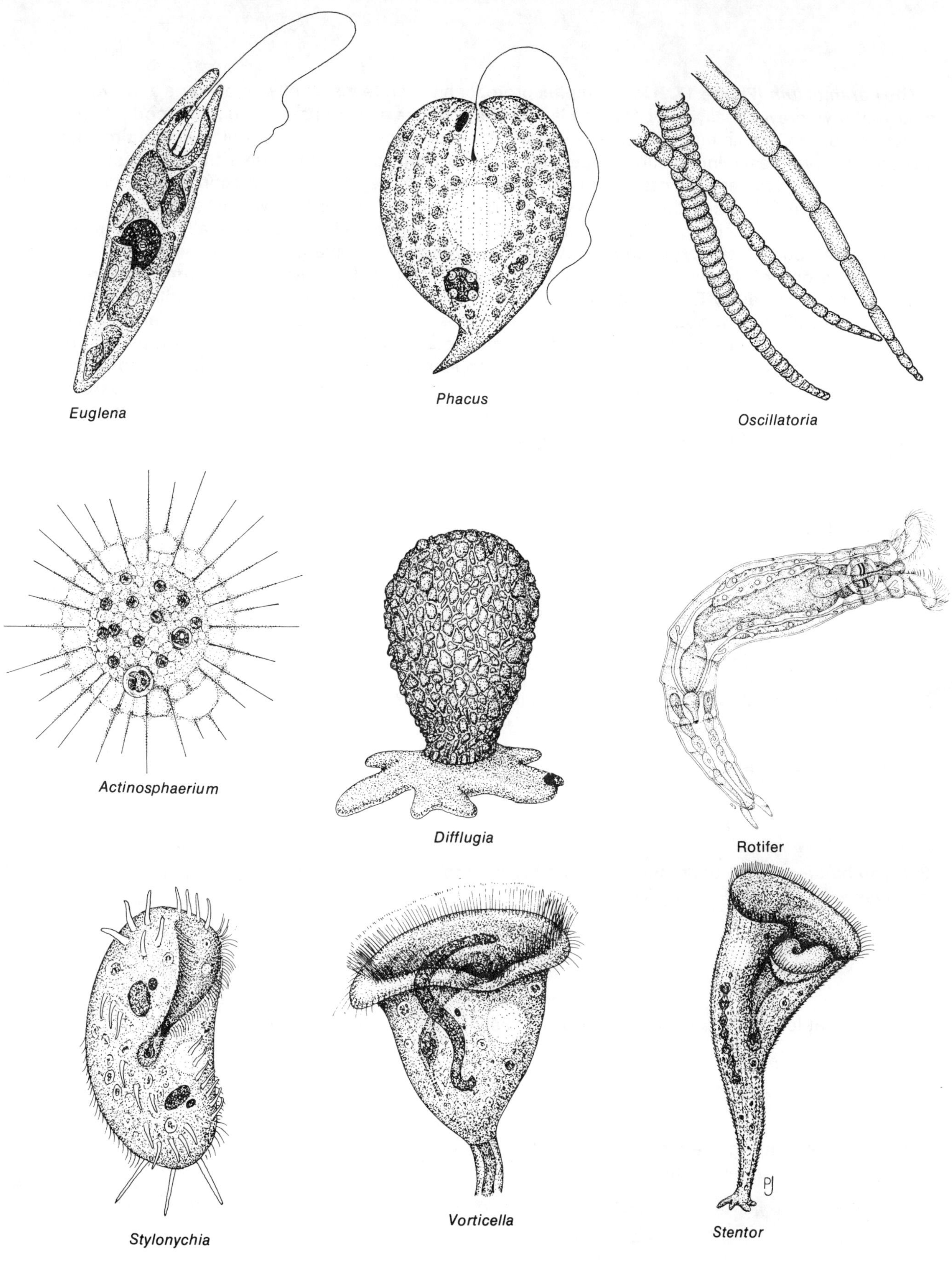

Figure 5.4. Some common free-living, freshwater organisms.

The *Paramecium* **(Figure 17.2)** is a complex protozoan which feeds chiefly on bacteria and lives in fresh water wherever decaying vegetation is present. Hair-like cilia on its surface carry food into the gullet and propel the animal through the water. Digestion occurs in food vacuoles which circulate through the cytoplasm. Indigestible materials are expelled through a pore near the posterior end. Two contractile vacuoles (one near either end of the organism) regulate its water content. *Paramecia* contain two types of nuclei: the larger macronucleus regulates cellular activity, while the micronucleus is involved in exchange of hereditary material during reproduction. Especially note the organism's method of locomotion and behavior when it encounters a large, solid object.

Euglena is one of the most common chlorophyll-containing flagellates, often forming a green scum on pond surfaces. Extending from the anterior end is a single whip-like flagellum, which propels the protozoan through the water. These organisms are also capable of a peculiar crawling motion. The orange-red eyespot functions in light detection. Numerous green chloroplasts are visible in the cytoplasm. The nucleus will probably appear as a clear area in the central cytoplasm.

Oscillatoria, a blue-green moneran alga, occurs in long filaments. Each individual cell is regarded as a whole organism, since they are all alike and self-sufficient. Chlorophyll and blue pigments are diffused throughout the cytoplasm. Frequently the filaments are seen to slide back and forth or oscillate gently, but the mechanism for this peculiar motion is not well understood. Since it is a moneran, no nucleus will be visible.

QUESTIONS

1. List cells you observed which contained storage areas or stored materials and give the type of materials contained.

2. What was the smallest living unit observed in today's lab?

   What was the largest?

3. If you had a drawing of an animal cell and wished to convert it to a drawing of a plant cell, what changes would you have to make?

4. Describe at least three structural similarities between plant and animal cells.

5. If you had a drawing of a moneran cell and wished to convert it to a drawing of a plant cell, what changes would you have to make?

6. How can you distinguish between the Golgi apparatus and the endoplasmic reticulum in an electron photomicrograph?

LABORATORY 6

THE CELL AS A UNIT OF FUNCTION — TISSUES AND ORGANS

OBJECTIVES: 1. To observe basic structural similarities in the cells of all tissues.
2. To observe the relationship between cell structure and cell function.

In the preceding laboratory, cells were studied as structural units. In multicellular plants and animals, a group of cells performing similar functions is a **tissue.** A group of tissues cooperating in the performance of a particular function is an **organ.** Tissue cells are often localized within a given area of the organism, but sometimes they are not. For example, blood is a tissue even though it pervades all parts of the living animal; epidermis is a tissue even though it covers the entire surface of the plant.

Tissue cells specialized to perform specific functions arise from unspecialized embryonic growth tissues. By the time the organism has matured, most cells have differentiated into various shapes and sizes and acquired organelles enabling them to carry out the specified task. A few adult cells retain the potential for further growth and specialization. Each tissue carries out a specific function, and each is dependent upon the others. Failure to perform any of these functions is harmful to the entire organism. This **division of labor** (assumption of a specific task by a tissue or organ) contributes to the continued existence of the organism as a whole. A tissue may be composed of only one, or of several types of cells. The particular amount of each tissue present and the arrangement of the various tissue types enable an organ to perform its specific function.

It is important to realize that throughout the biological world the structure (shape, size, organization, etc.) of a given tissue is correlated with the function it performs. Thus, one way of organizing a study of tissues is to consider the basic shapes and arrangements of the cells.

Typical cell shapes and their functions and arrangement are:

1. **Tubular** or spindle-shaped cells usually function in movement of materials, in support, or in muscular contraction.
2. **Flattened sheets** of cells act as linings or protective coverings.
3. **Layers of columnar cells** may function in secretion or absorption of materials.
4. **Irregular networks** of cells frequently function as connective or supporting tissues, or may be specialized for such functions as food storage or impulse transmission (nerve cells).
5. Relatively **small unspecialized cells** may function in growth. For example, the ring of cambium in the stem produces cells which differentiate into new areas of specialized tissue. This results in an increase in the diameter of the stem. In animals, undifferentiated regions occur in the skin, intestine and blood-forming organs, but are not as obvious as in plants. During embryonic development, adult tissues originate from unspecialized cells.

PLANT CELLS AND TISSUES

Sunflower *(Helianthus)* Stem

The sunflower is an example of a non-woody stem which lives only one growing season. Other stem types contain basically similar tissues, but frequently differ somewhat in their arrangement.

With low power examine both the longitudinal and the cross section of this stem on the prepared slide. First obtain an overall picture of tissue arrangement by moving the slide about and studying the entire section. Use the following description, and, if necessary, your text, to label **Figure 6.1. Begin your observations on the cross section first.** Then study the longitudinal section and add what you see to the diagram of the longitudinal section in **Figure 6.1.**

The following tissues will be observed, beginning at the outer edge of the stem and working toward the center:

Epidermis — a protective covering one cell layer in thickness.

Cortex — tissue composed of two types of cells. It extends from the epidermis to the ring of cambium, but does not include the vascular bundles.

1. **collenchyma** — cells have irregularly thickened walls and are supportive in function. They occur as a layer a few cells thick toward the outer edge of the cortex.
2. **parenchyma** — cells composing the remainder of the cortex. They function in food storage. Virtually all plant tissues contain these thin-walled, many-sided cells which compose the bulk of the sunflower stem.

Vascular bundles — These tissues, specialized for conduction, are oval in cross section, tubular in longitudinal section, and arranged in a circle. They contain several cell and tissue types.

1. **phloem fibers** — On the outer edge of each bundle is an obvious cluster of small, thick-walled, red-stained fibers. These cells are no longer living, but have very strong walls and support the stem and vascular bundle. Familiar examples include the "strings" in celery and linen, flax and hemp.
2. **phloem** — This green-stained tissue conducts dissolved food from the leaves to the roots. The larger cells are elongated and thin-walled, forming long tubes, with end walls perforated like a sieve, allowing dissolved food materials to pass from cell to cell. Mature phloem cells have lost their nuclei, and four smaller living **companion cells** usually surround each **sieve tube cell.**
3. **xylem** — These large, red-staining cells are elongated and lack end walls and protoplasm. They are arranged end-to-end, forming continuous tubes through which water and dissolved minerals move upward from the roots to the leaves. In the longitudinal section, it will be seen that walls of xylem tubes are generally thickened by rings or spirals of cellulose and may contain pits.

Cambium — separates the xylem from the phloem, forming a continuous ring around the stem. The thin-walled, brick-shaped cells form a layer three to five cells in thickness, usually stained green. The cambium retains the ability to divide, and produces new phloem cells to the outside, new xylem cells to the inside.

Pith — The center of the stem consists of large parenchymal cells (seen also in the cortex), many containing darkly stained grains of starch. The pith extends between each vascular bundle to the cambium.

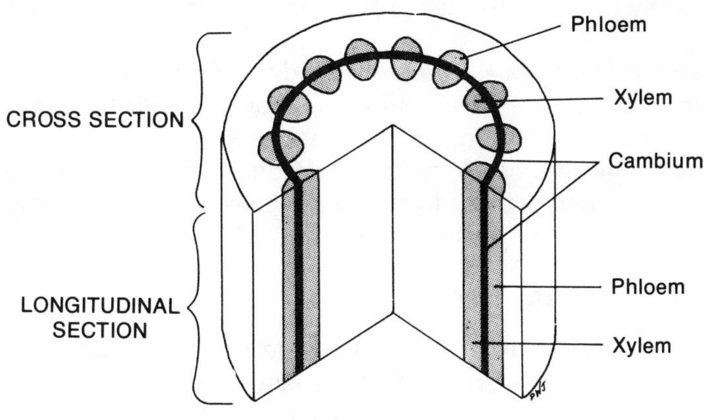

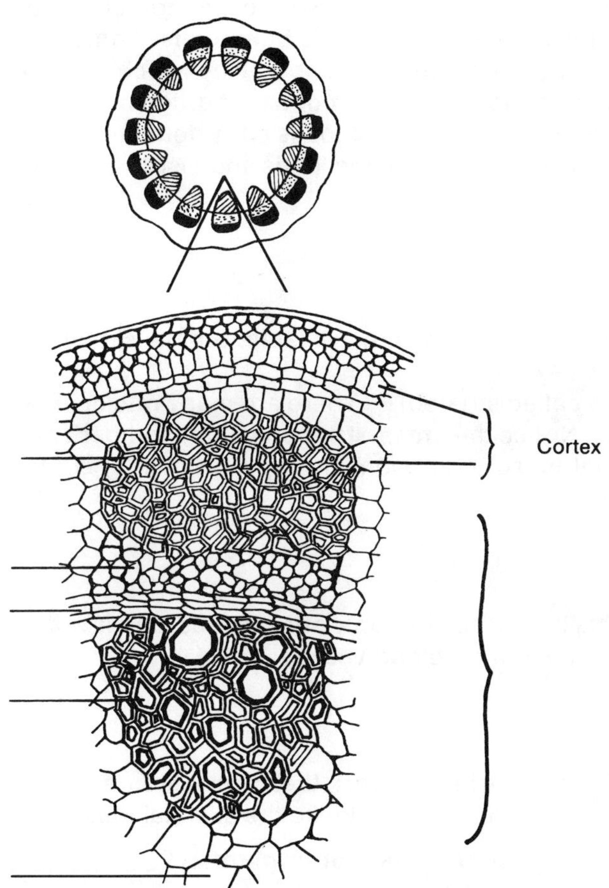

Figure 6.1. Sunflower stem (herbaceous dicot) showing orientation of tissues and detailed structure of a fibrovascular bundle.

ANIMAL CELLS AND TISSUES

Animal tissues are conveniently placed into four groups on the basis of similarity in structure and function:

1. **Epithelium** — closely packed layer of cells covering body surfaces or lining cavities or tubes.
2. **Connective** — (including vascular) — cells separated by a nonliving matrix; this tissue is usually involved in the support or connection of other tissues.
3. **Muscles** — elongated cells with the capacity for contraction in response to a stimulus.
4. **Nerve** — cells are "irritable"; they respond to a stimulus by a wave of excitation transmitted as a nerve impulse.

Cheek Cells (Example of Epithelial Tissue)

With the blunt end of a toothpick, **gently** scrape the inside of your cheek a few times. Stir the toothpick around in a small drop of iodine solution on a glass slide and add a coverslip. Observe under low power and find some unfolded cells. Notice their flat shape and darkly staining nuclei. Epithelial cells are shed regularly and replaced by new ones formed in lower layers.

Bone (Example of Connective Tissue) Demonstration

Study the slide and note a **matrix** of generally concentric rings around a canal. Each set of rings and canal is called a **Haversian system.** The matrix, secreted by the bone cells, is infiltrated with calcium to give the bone rigidity. Each Haversian canal contains a blood vessel and a nerve. The small, black, feathery areas in the rings are the **bone cells.** Compare the slide to **Figure 6.2.**

Bone, with its rigid composition, differs from cartilage which has a rubbery textured matrix. Other connective tissues, such as mesenteries, tendons, and ligaments, have fibrous matrices with the properties of flexibility and elasticity. The blood is considered to be a type of connective tissue. How does it compare with bone with respect to the kinds and distribution of cells and the type of matrix?

Muscles

Examples of the three muscle types are available. Simple sketches of each will be helpful when you start to review for an exam.

Skeletal muscle (also called striated) — There are many nuclei in one long, rope-like cell located near the cell membrane. Notice the **cross-striations** (light and dark bands perpendicular to the long axis of the cell). Skeletal muscles are consciously (voluntarily) controlled and are responsible for movements of the skeleton.

Smooth muscle — This type of cell has only one nucleus, and is not striated. It is involuntary, found on walls of the intestine and lining blood vessels.

Cardiac muscle — Found only in the heart, these cells are uniquely capable of rhythmic contractions. The cells show cross-striations, but unlike the skeletal muscle, they may be branched. Would you expect cardiac muscle to be voluntary or involuntary? ____________________

Neurons

Nerve cells and nervous tissue will be examined later in Laboratory 10.

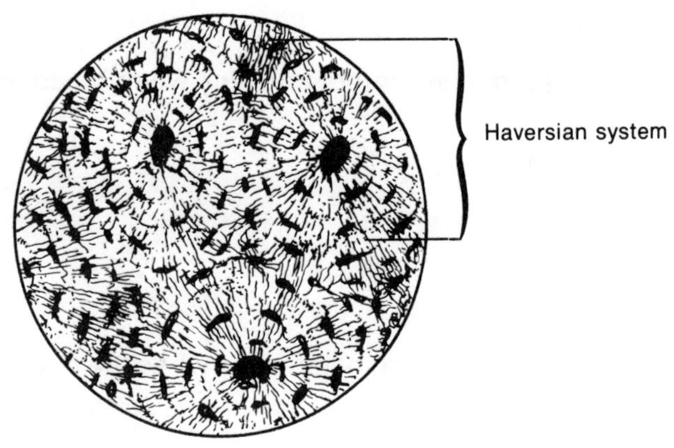

Figure 6.2. Cross section of bone.

CROSS SECTION OF A FROG'S SMALL INTESTINE (TISSUES IN AN ORGAN)

The digestive tract is composed of the same tissues, arranged in a similar order, in virtually all animals with backbones (fish, amphibians, reptiles, birds, and mammals). The digestive tube is covered externally and lined internally with epithelial tissues arranged in distinct, concentric layers. Muscle layers and connective tissues are found between the two epithelial layers.

The prepared slide should be examined first under low power to get an overall view of its layered structure. Complete the indicated labeling of **Figure 6.3** as you study the following. It would also be helpful if you drew a pie-shaped wedge in the space provided and labeled it. This drawing would allow you to record details not given in the diagram.

Beginning at the outermost edge, the intestinal wall is composed of four major coats or tissue layers.

Serosa — This is a single layer of flattened cells called **squamous epithelium.** Even using the high power objective, the nature of these cells is difficult to distinguish, but the darkly-stained nuclei are usually visible.

Muscularis — This tissue layer is composed almost entirely of spindle-shaped **smooth muscle cells** arranged in two directions. Rhythmic contractions of the muscularis produce a wavelike motion (peristalsis), moving the food along the intestinal tube. The cells of the outer **longitudinal muscle layer** appear circular in cross section. The cells of the thicker, inner **circular muscle layer** appear elongated since they are viewed from the side. Distinct cell membranes are not easily seen, but darkly-stained nuclei should be visible.

Submucosa — Composed chiefly of fibrous **connective tissue,** this tissue extends into the folds of the inner layer. Numerous darkly outlined open areas within the tissue are the blood and lymph vessels. Absorbed nutrients enter these vessels and are transported throughout the body. The submucosa appears lighter than the muscularis in the stained slides.

Mucosa — This innermost layer functions chiefly in absorption of dissolved nutrients, but also secretes a number of digestive juices and lubricants into the intestine. Its highly folded nature greatly increases total surface area for greater absorption. It consists of a single layer of **columnar epithelium** bordering the **lumen** (cavity) of the intestine. The large, darkly stained sphere within each cell is the nucleus.

Which has the greater surface area, the serosa or the mucosa? ______________________

Explain the functional significance of this difference in surface area.

What is the effect on the length of a section of intestine when the longitudinal smooth muscle layer contracts?

Effect on the diameter?

What is the effect on the length of a section of intestine when the circular smooth muscle contracts?

Effect on the diameter?

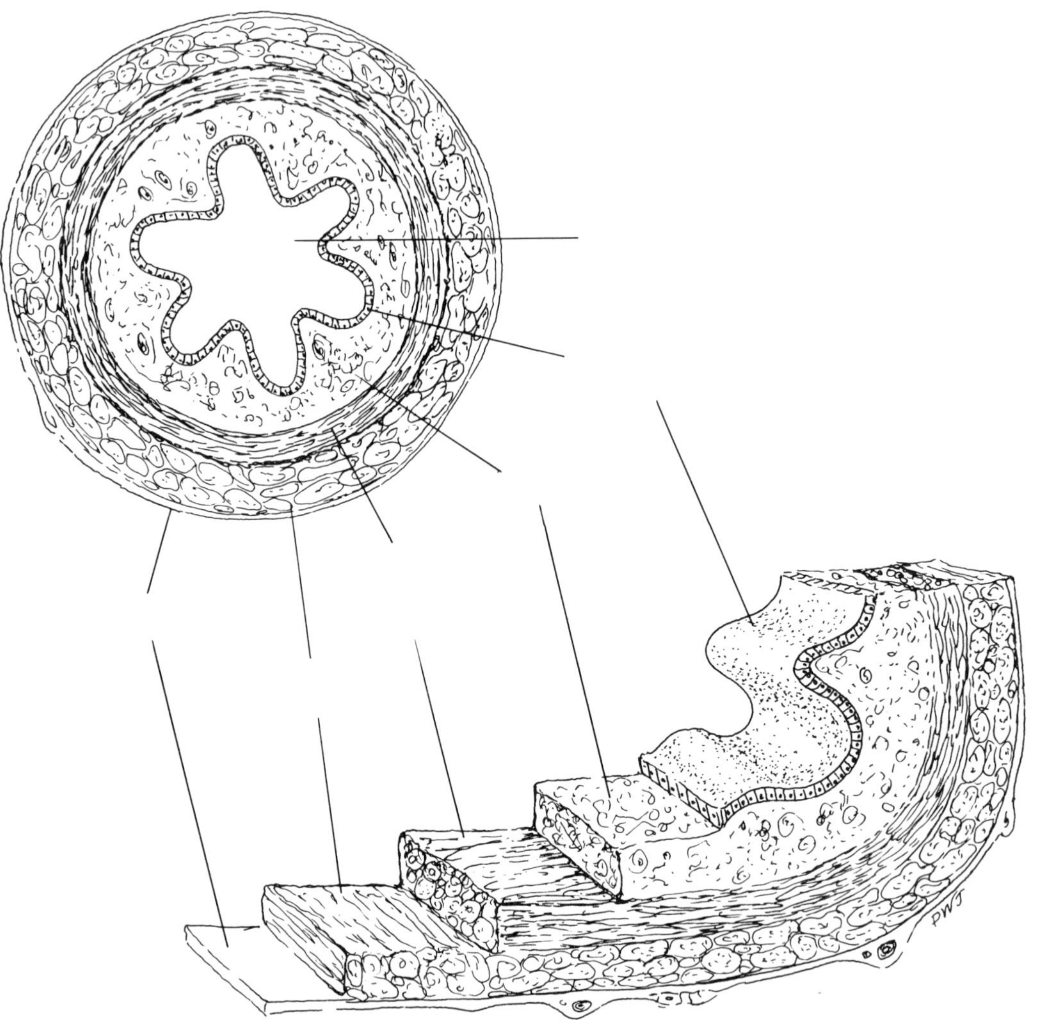

Figure 6.3. Diagram of tissues in the small intestine of the frog.

Draw and label a pie-shaped wedge of the small intestine below.

QUESTIONS

1. List the specific tissues of the following shapes which you studied and indicate their function.

   Tubular or spindle-shaped —

   Flattened sheets —

   Columnar —

   Irregular networks —

   Small, unspecialized —

2. List two general ways in which tissues of plants and animals are similar.

3. List two general ways in which tissues of plants and animals differ.

4. Does the similarity of structure between two tissues always imply similarity of function? List an example to support your answer.

5. Does similarity of function between two tissues always involve similarity of structure? List an example to support your answer.

LABORATORY 7
CELLULAR METABOLISM AND NUTRITION

OBJECTIVES: 1. To demonstrate some chemical processes associated with cellular respiration and photosynthesis.
2. To observe autotrophic and heterotrophic nutrition.

We have established that living organisms are composed of extremely complex arrangements of atoms and molecules. The maintenance of this highly ordered system requires a constant expenditure of energy. To obtain this energy and to use it as efficiently as possible, organisms utilize sequences of stepwise chemical reactions to alter chemical compounds obtained from the environment. All of the chemical pathways considered together compose the **metabolism** of the organism. These reactions do not manufacture energy, they can only transform energy in chemical compounds to a usable form.

The primary energy source for life is the sun. Green plants use the radiant energy of sunlight to build sugars out of carbon dioxide and water. They are capable of this remarkable activity because chlorophyll absorbs particular wavelengths of light and converts the energy into chemical bonds within the sugar molecule in a process called **photosynthesis.**

However, the energy is not immediately available for use by the cell until it can be released from the sugar molecule and passed on to promote other reactions. This occurs as sugar is broken down, usually back to carbon dioxide and water, in the presence of oxygen. Most of the released energy is put into the form of energy-rich phosphate bonds which can be stored and used as needed by the cell (such as in ATP, adenosine triphosphate). These reactions are common to both plant and animal cells and are known as **cellular respiration.** The major difference, then, between the metabolism of plants and animals is the ability of plants to "trap" sunlight energy through photosynthesis. Animals are therefore dependent either directly or indirectly on plants to provide them with energy-yielding fuel molecules, or "food."

RESPIRATION

One of the most important of several fuel molecules used by the cell is glucose. Using this sugar as an example, a much-simplified and general equation for cellular respiration may be written as follows:

$$C_6H_{12}O_6 + 6O_2 \longrightarrow 6CO_2 + 6H_2O + \text{Energy}$$

We will use germinating bean seeds to show that CO_2 is evolved as a product of respiration, as shown by the formula.

1. Work in groups of four.
2. Place about twenty germinating beans into a flask and stopper **tightly.** Repeat with dry beans in a second flask, also stoppered. Set aside for at least thirty minutes (preferably longer) while you proceed with other parts of the exercise.
3. After beans have been enclosed in the flasks for the recommended period, take five test tubes and half-fill them with tap water. Add 6-8 drops of 0.1% brom thymol blue to each. Brom thymol blue is an acid-base indicator (yellow at pH <6.0 and blue at pH >6). Leave one tube untreated as a negative control; add a single drop of H_2SO_4 (positive control) to tube #2. Pour water slowly through the delivery apparatus containing dry beans while tube #3 is held so that gases forced out of the flask bubble through the brom thymol blue **(Figure 7.1). Important:** Do not fill the flask over two-thirds full of water. If no color change is observed by this time, record no change.

4. The same procedure is followed for tube #4 using the apparatus containing germinating beans. To check the reliability of the test for CO_2, one person should blow into tube #5 using a straw or pipette.
If CO_2 is produced, then:
$CO_2 + H_2O \longrightarrow H_2CO_3$ (carbonic acid)
The acid will be detectable with the change in color of the indicator.
5. Record your results:

Tube No.	Material added to H_2O and Brom Thymol Blue	Results
1	None (negative control)	
2	Sulfuric Acid (positive control)	
3	Gas from dry beans	
4	Gas from germinating beans	
5	Breath	

6. Now remove the stopper from the delivery flask containing germinating beans. Carefully light a match and insert into the flask. Note the time required to extinguish the flame. Repeat this procedure using the flask with the dry beans. Again note time required. Record your results and your interpretation.

7. Return the germinating beans to the original container and the dry (now wet) beans to the designated container.

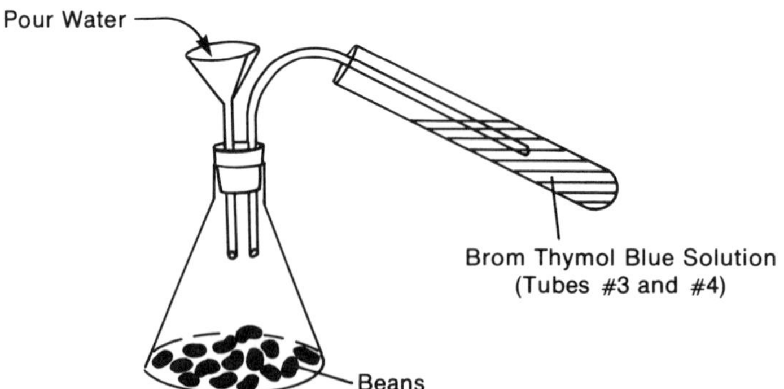

Figure 7.1. Apparatus for demonstrating CO_2 as an end-product of respiration.

QUESTIONS:

1. Change to a yellow color from a blue with brom thymol blue is a positive test for what substance?

2. In your experiment, the indicator showed that__________ is a product of respiration.
3. Define (using help from your text, if needed):
   a. aerobic respiration —

   b. anaerobic respiration —

   c. fermentation —

   d. glycolysis —

   e. cytochrome —

   f. electron transport chain —

NUTRITION

AUTOTROPHIC NUTRITION

Autotrophic organisms utilize simple, inorganic materials (such as CO_2 and H_2O) to synthesize complex, energy-rich organic molecules for their own use. (Heterotrophic organisms will also directly or indirectly use these molecules.) The original source of energy for life is sunlight. In the process of **photosynthesis,** energy from the sun is converted and trapped in organic molecules (carbohydrates) synthesized from CO_2 and H_2O. Light energy is required for the initial reaction (light reaction), but other steps proceed in the absence of light (dark reactions). Most photosynthetic organisms convert light energy through use of the pigment chlorophyll contained in chloroplasts. Chlorophyll traps light energy and uses the energy to split water molecules. The reaction, known as **photolysis,** is shown below.

$$2\ H_2O + \text{Light} \longrightarrow O_2 + 2H_2$$

Hydrogens released in photolysis may combine with CO_2 (reduction) to produce (CH_2O), a unit which may be used as an energy source or for building more complex carbohydrates. The hydrogens may combine with oxygen to produce water. Thus, light, carbon dioxide and chlorophyll are required for the process of photosynthesis. A simplified form of the overall reaction is shown below.

$$6\ CO_2 + 12\ H_2O^* \xrightarrow[\text{Chloroplasts}]{\text{Light Energy}} (CH_2O)_6 + 6O_2^* + 6H_2O$$

The asterisk indicates a radioactive isotope of oxygen which was used experimentally to demonstrate that the free oxygen released was originally a part of the water molecule. The unit (CH_2O) may be utilized by the plant for an energy source in the form of a 3-carbon unit or a 6-carbon unit, or it may be converted to the polysaccharide starch (a long chain of 6-carbon units), and stored.

IMPORTANT FACTORS IN PHOTOSYNTHESIS

Chlorophyll (Demonstration)

1. Leaves from a variegated coleus (or a similar plant) kept under a strong light for 48 hours will be boiled first in water (to kill tissues and remove water soluble pigments) and then in alcohol (to extract chlorophyll). The leaves will then be tested for the presence of starch by adding a dilute iodine solution. Areas containing starch will turn a purplish-brown color. The presence of starch may be used as a measurement of excess photosynthesis, leading to production of a storage form of carbohydrate.
2. Alternatively, or in conjunction with the coleus, a plant (such as a geranium) may be placed in the dark for 24 hours. It is then removed and a piece of paper used to cover a portion of the leaf. The plant is then placed in a strong light three hours to induce excess photosynthesis. This leaf may then be tested for starch.
3. Sketch the appearance of the leaves before boiling and after completion of the test, indicating colors.

4. Conclusions:

Carbon dioxide (CO_2) and Light (Demonstration)

1. The instructor will add 0.1% brom thymol blue to a flask of water. Sufficient indicator will be added to produce a dark blue color.
2. Four test tubes will be labeled A-D.
3. Tube A will be filled with the blue colored solution (control).
4. The remainder of the solution will have CO_2 bubbled through it (by exhaling through a straw or a pipette) until the color just changes to yellow. The indicator will change to yellow at an acid pH. Color change, therefore, may be used as an indirect indicator of the amount of CO_2 present.
5. Tubes B, C, and D will be filled with the yellow solution, and Tube B will remain as a control for this color.
6. *Elodea* will be placed in tubes C and D and a piece of aluminum foil wrapped around C.
7. All four tubes will be stoppered and placed in front of a strong light source.
8. Observe the tubes at 10-minute intervals for color change. If CO_2 is utilized by the *Elodea,* the pH should change and the color will become blue.

9. Results:

Tube	Initial Color	Final Color	Time Required
A	Blue	________	________
B	Yellow	________	________
C	________	________	________
D	________	________	________

10. Conclusions:

Oxygen — a by-product of photosynthesis (Students work in pairs)

1. Fill a tube approximately half-full with 0.4% $NaHCO_3$. This acts as a source of CO_2, so that CO_2 will not be a limiting factor in this experiment.
2. Obtain a piece of *Elodea* approximately 5 cm long and strip a few leaves from one end. Cut the end of the stem diagonally using a scalpel. Avoid touching the cut surface.
3. Invert the sprig and push it into a tube with the cut end up until the end of the stem is approximately one inch below the surface. Do not allow the *Elodea* to remain out of water any longer than necessary.
4. Place the tube in front of a strong light until bubble formation begins. Count the number of bubbles released per minute.
5. The importance of light intensity in oxygen production may also be shown using this same set-up. Simply vary the distance to the light source and repeat the counting of bubbles.
6. Results:

Bright Light No. bubbles/min.	Dim Light No. bubbles/min.
________	________

7. Conclusions:

8. You have been told that the bubbles are oxygen — can you prove this?

QUESTIONS:

1. Define autotroph.

2. Define photolysis.

3. How did you show the need for chlorophyll in photosynthesis?

4. How did you show the need for light in photosynthesis?

5. How did you show that CO_2 is used in photosynthesis?

6. What effect would you anticipate if the tubes containing brom thymol blue and *Elodea* were placed at a temperature of 15° C during the experiment?

7. What changes in the number of bubbles of oxygen evolved/min by the *Elodea* would you expect to find if the tubes in the last part of the exercise were placed in the bright light at 37°C? in dim light at 37° C?

HETEROTROPHIC NUTRITION

Heterotrophs require preformed organic compounds to furnish their needs for both matter and energy, as they lack the ability to synthesize food from inorganic materials. Organic molecules used consist of one or more of the following: amino acids, sugars, fatty acids, and glycerol. These molecules are usually obtained from more complex substances which are broken down in the process of digestion. A variety of methods for obtaining these molecules and converting them to usable forms have been evolved by living organisms. Most heterotrophic organisms (which includes all animals and fungi and most bacteria) may be classified under one of the following types.

1. **Saprophytic nutrition.** These organisms live on dead organic material, absorbing required nutrients through their cell membranes. They lack structures for taking in large food particles. Secreted enzymes diffuse into the environment and convert complex molecules to smaller molecules which are able to diffuse into the cell. The common black bread mold *(Rhizopus nigricans)* is an example.
2. **Parasitic nutrition.** These organisms obtain their nourishment from the body of the host (the living plant or animal on which they reside) by ingesting and absorbing organic particles or by absorption of organic molecules through their cell membranes. Parasites usually cause some degree of harm to the host. Mistletoe may be classified as both a parasite and an autotroph because it obtains food both ways.
3. **Holozoic nutrition.** This type of heterotroph obtains food as solid particles which are eaten, digested, and absorbed. A variety of sensory, nervous, muscular, and digestive methods and structures have evolved in order to pursue, capture, and ingest solid particles.

Although heterotrophic organisms utilize a variety of means for obtaining food, the process of converting complex molecules into usable ones is remarkably uniform. Such widespread uniformity in the digestive process should not be particularly surprising, since the bodies of all organisms are composed of the same basic types of organic compounds (carbohydrates, lipids, proteins, and nucleic acids), and chemical breakdown results in the same basic subunits regardless of the organism involved. These **end-products of digestion** — amino acids, sugars, fatty acids, and glycerol — are absorbed by the individual cell or cells of the organism, and used to build new cell substances or burned in cellular respiration.

In each case, digestive breakdown is accomplished by specialized digestive enzymes which break bonds between the molecular subunits by adding a molecule of water **(hydrolysis).** Some of these enzymes, usually the first ones involved, function most efficiently in an acid environment; others, which complete the digestive process, require alkaline conditions to function.

Just as specialized means of ingesting foods occur among heterotrophs, there are also regions in these organisms specialized for digestion. The digestive processes may all occur in one location, such as the food vacuole of single-celled animals like *Paramecium,* or in many specialized subregions of a complex digestive system, as in the earthworm or in man. In each case, ingested food is first broken into small bits to increase the surface area exposed to digestion, and the material is mixed with digestive enzymes in an acid digestive fluid. Where specialization is present, this occurs in the anterior region of the digestive tube — the stomach, for instance, in man. The partially digested material is then subjected to the action of further digestive enzymes in an alkaline digestive fluid. This is either secreted into the same space, as in the food vacuoles of protozoa, or is accomplished by moving the materials to another, usually posterior, digestive region — the intestine. This movement occurs as a result of rhythmic waves of alternate contraction and relaxation of smooth muscle layers in the intestinal walls, and is called **peristalsis.** Digested molecules are then absorbed either actively or by diffusion and distributed to all regions of the body. In some cases, digestive systems may combine partial extracellular breakdown in the lumen or cavity of the digestive system with an intracellular final digestion in food vacuoles of digestive cells. There are also several modifications which occur in digestive systems which increase surface area, and thereby provide greater absorptive capacity. Two of these modifications are illustrated by the intestinal wall structure of the frog and the earthworm. There may also be pockets or branches of the intestinal walls where food is held for relatively long periods of time so that it may be more completely utilized.

HOLOZOIC NUTRITION IN A SINGLE CELLED ORGANISM

1. Place a small drop of *Paramecia* culture or pond water on a slide and add a small drop of methyl cellulose, which will cause the animals to swim slowly.
2. Mix the solution of yeast stained with Congo red. Dip a toothpick approximately 0.5 cm into the solution, remove and touch against edge of container to remove excess. Touch stained end of toothpick to drop of water and remove immediately. **Avoid excess yeast.**
3. Use **opposite** end of toothpick to mix yeast into the drop of water and add a coverslip.
4. Locate a large *Paramecium* with low power, then switch to high power. Observe the various organelles. Observe how yeast are ingested, the formation and movement of a food vacuole and ejection of undigested materials. Also, look for color changes within the vacuole. (The indicator dye will turn blue at pH 3.)
5. If you used pond water, look around the slide and observe other organisms for possible ingestion of yeast.
6. Make an outline drawing of a *Paramecium* showing ingested yeast. If you were able to observe movement of a food vacuole, indicate this with dotted lines on your drawing.

HOLOZOIC NUTRITION IN A METAZOAN

In the one-celled animals just observed, digestive organelles are called food vacuoles. The increasing complexity found in other groups has led to numerous alterations in this simple system.

1. **The water flea** (Film loop showing feeding process in *Daphnia.)*
Daphnia, the water flea, is a small animal related to shrimp, lobsters, and crabs. They may be found in a wide variety of fresh water habitats ranging from puddles to open lakes.

Daphnia **(Figure 7.2)** feeds by using its trunk appendages to create a water current through the space between the two rows of these paired appendages. Filtering setae collect food particles from the water and pass them to a central food groove leading to the mouth opening.

The digestive system, clearly visible through the transparent carapace or shell, consists of a mouth surrounded by jaws, and a long intestine terminating in an anal opening at the end of the abdomen. The midregion of the intestine approximately below the heart is where most digestion occurs. Waves of peristalsis carry partially digested food back through the intestine to the anus, or forward into the two blind pouches, the midgut ceca. Absorption of digested food materials occurs all along the midgut and through the walls of the midgut ceca.

2. **The Earthworm**

This group is characterized by a **complete digestive tract** with differentiation of various areas. The earthworm's digestive tract is diagrammatically shown in longitudinal cross section **(Figure 7.3a).** This animal demonstrates the **tube-within-a-tube** body plan.

Using low power, observe the slide showing a transverse cross section of this animal.

Very diagrammatically, the general plan of this section is shown in **Figure 7.3b.** Use this diagram as a reference to aid in locating areas and structures on your slide. As you compare drawing and slide you might want to add additional detail to this figure.

The **intestine** is located in the center of the **coelom.** Note the indentation or invagination (known as the **typhlosole)** in the dorsal wall of the intestine. The open space in the center of the intestine is called the **lumen.**

Switch to high power and observe the cellular detail of the body and intestinal walls. Note the two **layers of muscles (circular** and **longitudinal)** which contract to shorten and narrow the lumen, thereby pushing materials through the intestine. Observe the layer of **epithelial tissue** covering the outside and inside of the intestine. The yellowish-colored **chloragen cells** surrounding the intestine synthesize and store glycogen and fat (functions performed by the liver in more advanced animals). Enzyme-secreting cells and absorptive cells compose the epithelium.

3. **The Frog**

Refer to **Figure 6.3.** Note the similarity in arrangement and structure of the frog and earthworm intestine, even though the two animals are taxonomically rather distantly related.

Compare the frog and earthworm intestine.

Intestine	**Earthworm**	**Frog**
Shape	____________	____________
Muscles		
Layers (Number of)	____________	____________
Layers (Arrangement)	____________	____________

QUESTIONS:

1. Define:

   host —

   typhlosole —

   heterotroph —

   peristalsis —

2. What device increases the absorptive area in the frog intestine? in the earthworm?

3. Distinguish between holozoic and parasitic nutrition.

4. Distinguish between parasitic and saprophytic nutrition.

5. Distinguish between ingestion and digestion.

6. How is digestion in *Daphnia* typical of heterotrophic nutrition?

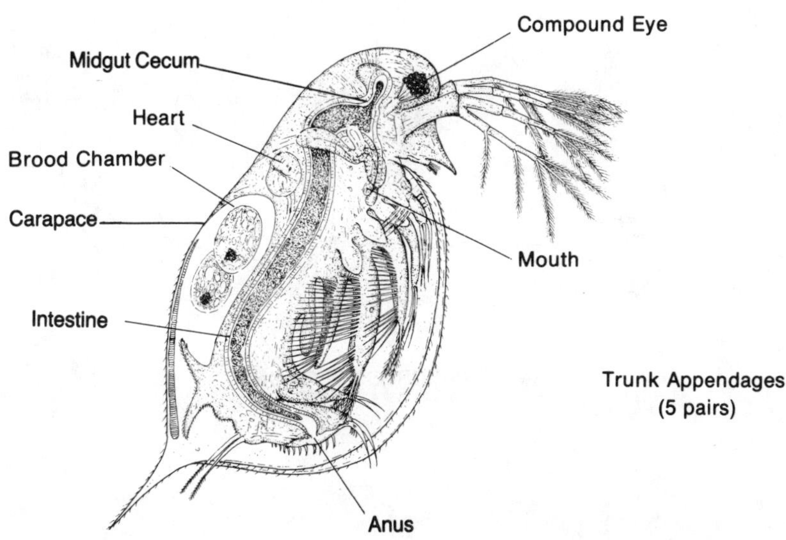

Figure 7.2. *Daphnia.*

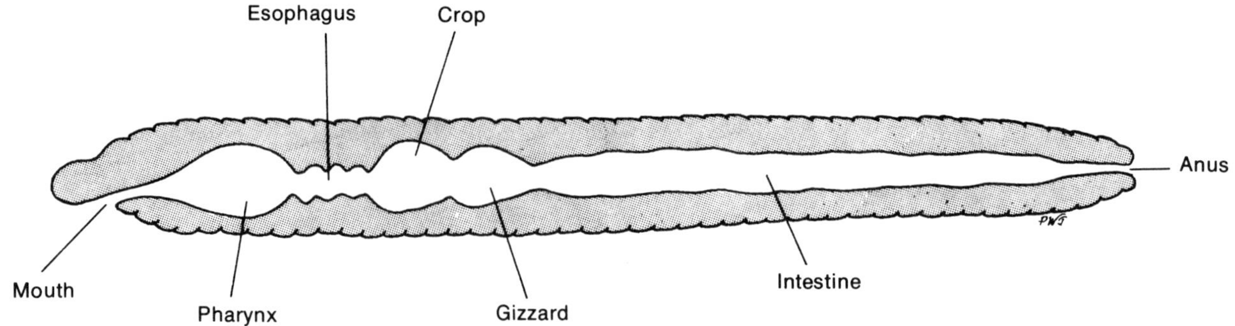

a. LONGITUDINAL CROSS SECTION

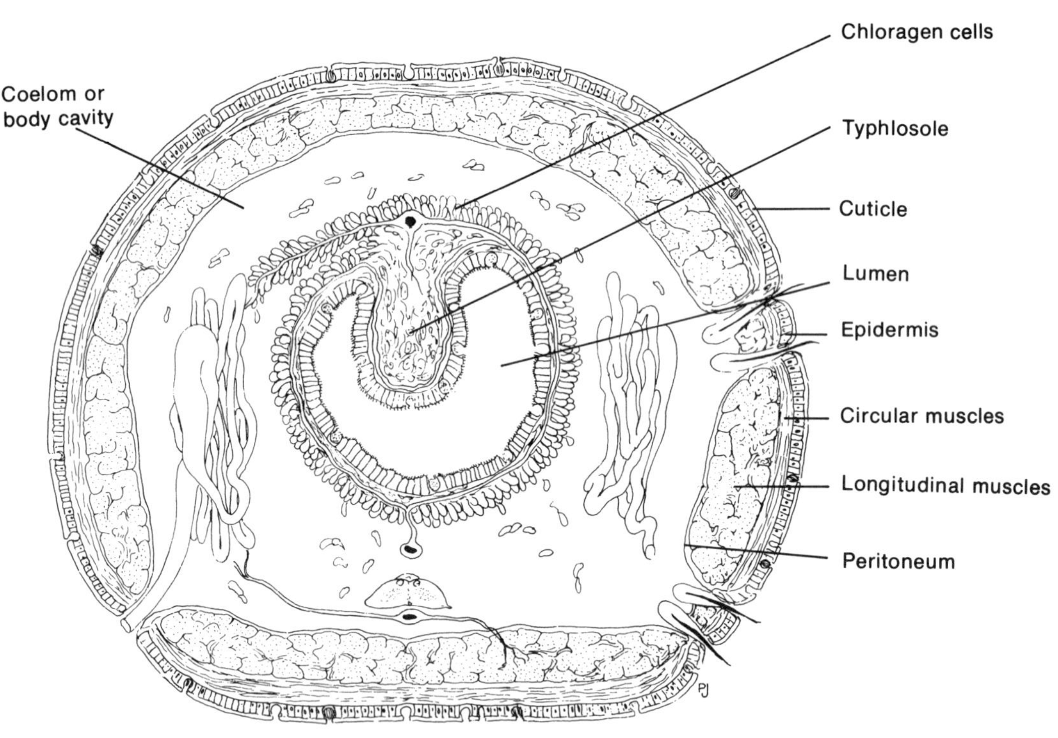

b. TRANSVERSE CROSS SECTION

Figure 7.3. Diagrammatic cross sections of earthworm. (Two pairs of setae on left side have been omitted from transverse section.)

LABORATORY 8
TRANSPORT SYSTEMS AND CIRCULATION

OBJECTIVES: 1. To study basic circulatory mechanisms and some specialized circulatory tissues.
2. To study the effect of exercise and caffeine on heart rate.

Each cell of an organism must have food, oxygen, and water, and must dispose of various metabolic products (basically nitrogenous wastes and carbon dioxide). In microscopic organisms and those with very thin body walls, such as sponges or jellyfish, simple diffusion from or into the fluid surrounding the organism may be sufficient **(Figure 8.1A and B).** Larger animals, however, may have very thick body walls, and internal cells are buried far too deeply to obtain materials directly from the surface. Hence, some system of transport is necessary to carry food from the digestive organs and oxygen from the surface or respiratory organs to the body cells, as well as transport carbon dioxide and various waste products of metabolism out of the organism.

Fluid in the circulatory system is also involved with maintaining the water balance of the organism. This aspect of circulatory function and the general circulatory mechanisms of plants will be discussed in the exercise on water balance. Additionally, in many higher animals, the circulatory system carries hormones or chemical messages from one organ of the body to another, and functions to help protect the body from invasion by disease-producing organisms.

Animals with incomplete or **open circulatory systems** (such as insects) send circulatory fluids from the heart into vessels which empty into irregular cavities and return to the heart via body spaces **(Figure 8.1C).** In organisms such as yourself, heavy-walled, muscular blood vessels **(arteries)** carry blood away from the heart to body cells. Arteries branch, becoming gradually smaller, until they end in a system of **capillaries (closed circulatory system)** with walls generally only one cell layer thick **(Figure 8.1D).**

The capillary bed or the open space (hemocoel) is the site of the actual exchange of food, oxygen, carbon dioxide and metabolic wastes which occurs between the circulatory fluid and the body cells. Blood is collected by thinner walled **veins** (containing valves to prevent back flow) and returned to the heart. This muscular pump may consist of one or several chambers, and serves to propel the circulatory fluid under pressure to the arteries.

In organisms with a vascular circulatory system, the circulatory fluid is called **blood.** The liquid part of the blood, the **plasma,** usually contains specialized respiratory pigments which carry oxygen and carbon dioxide. Most animals have a reddish colored pigment called **hemoglobin.** Other animals have differently colored pigments, such as hemocyanin which is blue. In insects, clams, snails, and earthworms, these pigments are dissolved in the plasma, but in vertebrates, the pigments are contained in specialized cells, **erythrocytes** or red blood cells. Additionally, vertebrate blood contains a second basic cell type, the **leucocyte** or white blood cell. Several different types of leucocytes may be recognized by differences in their appearance and function.

The rate at which the muscular heart contracts determines to a great extent the velocity with which fluid is pumped through the organism. While heart muscle is innately capable of contraction, the rate of contraction in a living organism is basically controlled by the organism's metabolic rate and oxygen requirements, its nervous system, and in animals such as man, by hormones circulating in the blood.

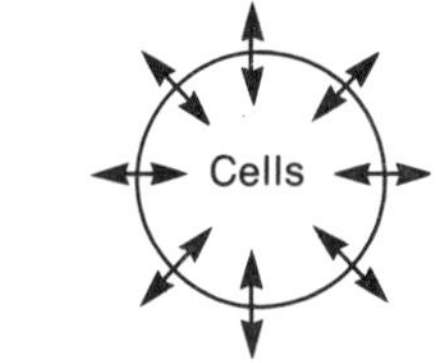

A. DIRECT EXCHANGE BETWEEN CELLS AND ENVIRONMENT

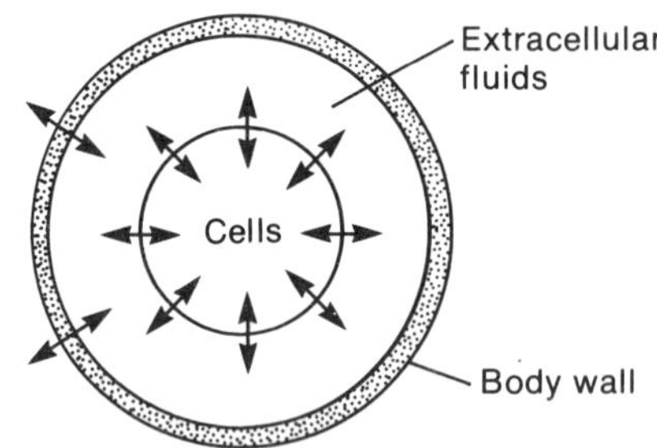

B. CIRCULATION OF EXTRACELLULAR FLUIDS IN THE ABSENCE OF A DEFINED CIRCULATORY SYSTEM

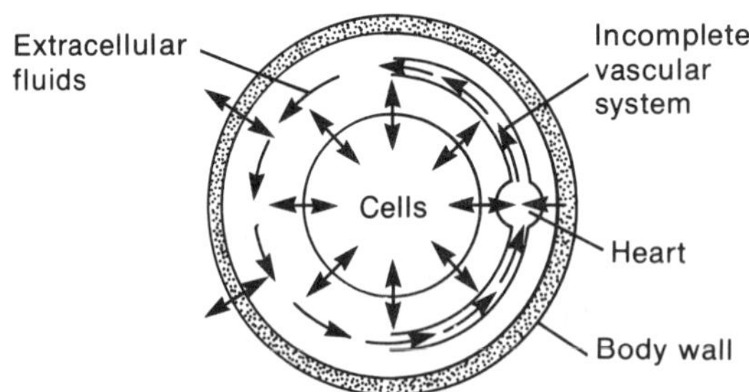

C. OPEN CIRCULATORY SYSTEM

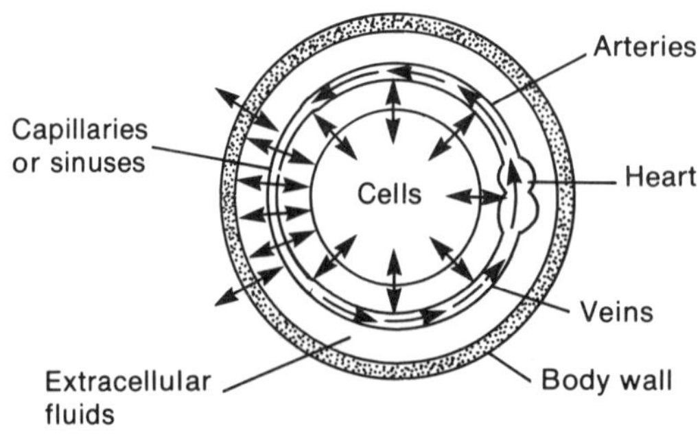

D. CLOSED CIRCULATORY SYSTEM

Figure 8.1. Types of circulatory and exchange systems.

COMPARATIVE STUDY OF VERTEBRATE HEARTS

The vascular system has undergone some striking changes during the evolution of vertebrates. Most of these changes are correlated with the shift from gills to lungs as the site of external respiration and with the development of an efficient, high pressure circulatory system. These changes occurred during the transition from water to land, and are necessary for active, terrestrial vertebrates.

Propulsive force is furnished by heart muscles. The vertebrate heart contains at least two chambers connected by valves which allow blood to flow in only one direction. Vessels transport the blood to and away from the heart as described previously.

The following structures may be observed on the various models and/or specimens. As you read the descriptions locate the structure on **Figure 8.2** and add labels to that figure where indicated.

Chambers

Sinus venosus — the large sac adjacent to the atria which receives venous blood and transmits it to the atria. Not present in higher vertebrates.

Atrium or **auricle** — anteriorly located chamber. May be divided into right and left atria.

Ventricle — posteriorly located, cone shaped region. May be divided by a **septum** into right and left ventricles.

Conus arteriosus — a sac posterior to the ventricle which transmits blood into the aorta. In higher vertebrates is fused with the ventricles and is not apparent.

Valves

Tricuspid — right atrioventricular valve.

Bicuspid (mitral) — left atrioventricular valve.

Pulmonary semilunar — base of pulmonary artery.

Aortic semilunar — base of aorta.

Vessels

Vena cava — (superior and inferior branches) major veins which collect blood from the body and empty into the sinus venosus or right atrium.

Pulmonary veins — carry oxygen-rich blood from lungs to heart, emptying into the left atrium.

Aorta — large, main artery which carries oxygenated blood from left ventricle or conus arteriosus to the body (its branches supply all parts of the body except the lungs).

Pulmonary arteries — carry oxygen-poor blood from right ventricle to lungs.

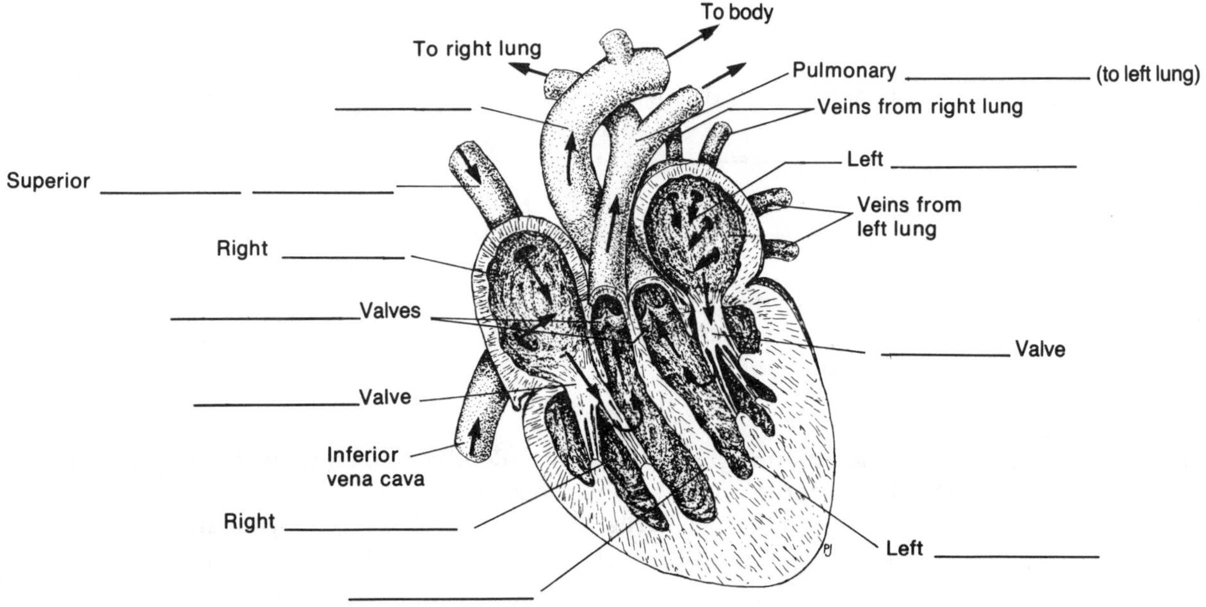

Figure 8.2. Blood flow through the human heart.

Group		Sinus Venosus	Single Atrium	Separate Right & Left Atria	Single Ventricle	Partially Divided Ventricle	Separate Right & Left Ventricles	Conus Arteriosu
FISH								
AMPHIBIAN (FROG)								
REPTILE	TURTLE							
	CROCODILE							
BIRD								
MAMMAL	DOG							
	HUMAN							

1. Study models and specimens of the hearts of various vertebrate animals. Identify chambers, valves, and major vessels. Compare the heart structures in the chart. Make a check mark for each structure found in the various vertebrate hearts.

2. Determine general trends in the evolution of the vertebrate heart with relation to:

number of chambers (not counting sinus venosus or conus arteriosus) ______________

presence of collecting regions (sinus venosus, conus arteriosus). ______________

3. On any of the models, examine the thickness of the atrial wall and compare with the ventricles.

Which have thicker walls? ______________

Why would this be an advantage?

4. Now compare the thickness of the walls of the right and left ventricles on one of the mammalian hearts. Which is thicker? ______________

Why?

5. In any of the models, look for extensions of tissue between the atria and the ventricles which act as valves. What is their function, and how do they work?

6. Trace an imaginary drop of blood through the fish heart, identifying all structures through which the blood passes. Repeat with the turtle and the human hearts.

Fish **Turtle** **Human**

7. The arrangement of heart chambers results in different pathways of blood as it circulates through the body.

Fish blood is pumped from the ventricle to the gill region, where oxygen diffuses into the blood and CO_2 is eliminated. Friction of the blood against capillary walls is great in this region, thus the blood rapidly loses pressure and leaves the gills by way of the dorsal aorta at low speed and pressure. Circulation through the rest of the body is sluggish, a factor not conducive to great activity. Draw arrows to indicate the pathway of blood in the fish.

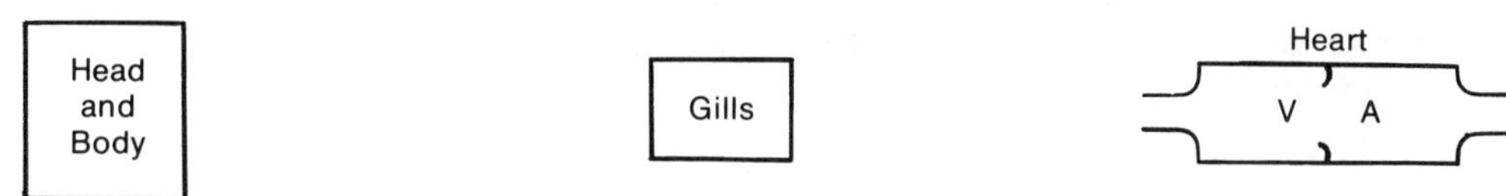

Why is circulation in the fish called "single circulation" while other vertebrates have "double circulation"?

Birds and mammals depend on lungs for gas exchange. The four-chambered heart really works as two halves. Blood moving to the lungs is under great pressure, having just left the right side of the heart. Blood moving into the body circulation, by way of the dorsal aorta, is also under much pressure, having just been forced from the left side of the heart. Complete separation of right and left sides of the heart also eliminates the inefficient mixing of oxygenated blood from the lungs and deoxygenated blood from the body. Complete the diagram with arrows indicating blood routes in the bird or mammal.

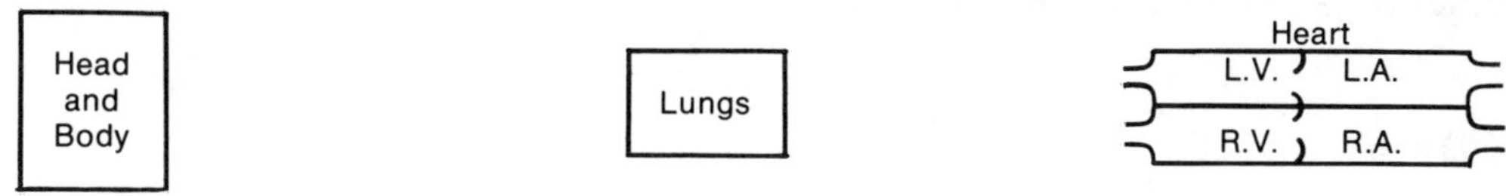

Reptiles and amphibians exhibit a situation intermediate between fish and the birds and mammals. The three-chambered heart (only one ventricle) only partially separates oxygenated and deoxygenated blood. Mammals and birds, due to efficient circulation, can maintain a constant body temperature and are called "warm-blooded" or **homeothermic.** This depends on a high metabolic rate and requires large, constant supplies of oxygen and food. Amphibians and reptiles are incapable of supplying these requirements and are "cold-blooded" or **poikilothermic** (have variable body temperatures). Indicate pathways of blood in an amphibian or reptile.

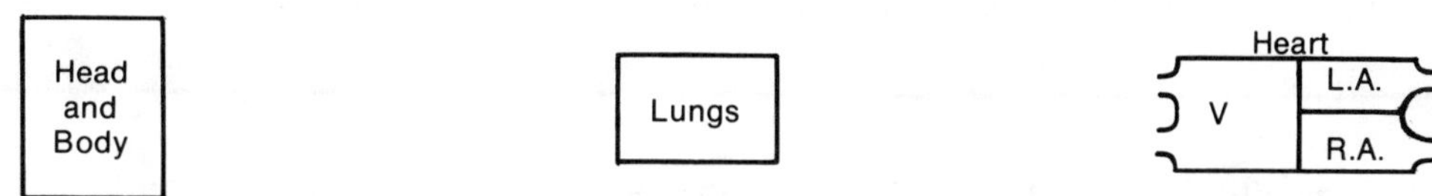

FACTORS ALTERING HEART RATE

Exercise

1. Obtain a measurement of your heart rate by having your partner measure your pulse for fifteen seconds after you have been sitting **quietly** for five minutes. The pulse rate may be felt in the wrist by placing your fingers firmly across it just below the heel of your hand.

Your resting pulse (beats/15 seconds) ______________________

Your resting pulse (beats/minute) ______________________

2. Walk quickly and quietly up two flights of stairs and back down again. Have partner measure your pulse rate immediately. You should then measure your partner's pulse rate before and after he or she climbs the two flights of stairs.

Your pulse after exertion (beats/15 seconds) ______________________

Your pulse after exertion (beats/minute) ______________________

What is the effect of exertion (stair climbing) on heart rate? ______________________

What causes this effect? ______________________

Chemicals

Select one student at your lab table to drink a cup of strong coffee. (Preferably the one who normally drinks the least coffee.) This student should drink the coffee quickly. Measure the pulse rate ten minutes afterwards. He or she should remain relatively inactive during this ten-minute period.

Your pulse after ingestion of caffeine (coffee)__________(where applicable)

What is the effect of caffeine on human heart rate? ______________________

What other organ systems do you think are affected by coffee? ______________________

Class Data

Post the data requested on the chalkboard as soon as you have finished these measurements. The instructor will average the data of the first fifteen students reporting, and post this information. Record these averages. All students will remain until class averages are complete.

Class averages (beats/minute) ______________________

Average resting pulse ______________________

Average pulse following exertion ______________________

Average resting pulse of individuals ingesting caffeine ______________________

Average pulse after ingestion of caffeine ______________________

What is the purpose of making several measurements of heartbeat rate and obtaining an average?

____

Were some individuals more affected by exertion than others?__________ Why?

HUMAN BLOOD CELLS

Observe a prepared slide of human blood cells. This was made by smearing a thin film of blood on the slide and then staining the cells. Focus the slide under low power and then switch to high power for observation. Locate several leucocytes and look for differences between them. Draw representative erythrocytes and leucocytes, showing their shape and relative size.

Which type of blood cell is more numerous on the slide?____________________ Do circulating human erythrocytes usually contain a nucleus?__________ In what way is the shape of red blood cells related to their function?____

____

In what way is the form and mobility of the leucocytes related to one of their protective functions?

____

TRANSPORT SYSTEMS AND CIRCULATION

Across

1. your circulatory fluid (backward)
3. major artery leading from heart to body
9. The____and the amphibian have a three-chambered heart.
10. At the base of the pulmonary artery is the semi____valve.
11. Fish have a____-chambered heart. (backward)
14. ____cuspid or right atrioventricular valve
15. carry deoxygenated blood to the heart
16. You have a closed____system.
18. erythrocytes or____blood cells
20. The____venosus is not present in higher vertebrates.
21. oxygen carrying pigment
22. movable structure between atrium and ventricle
24. Heart rate was determined by measuring the____.
26. ____h blood is pumped from the ventricle to the gills.
27. ____cuspid or left atrioventricular valve
28. a biconcave disc or erythro_____
29. The____cyte contains hemoglobin.
30. The bird and the____I depend on lungs for gas exchange.
31. The tri____valve separates the right atrium and the right ventricle in your heart.

Down

2. The mammal and the____heart have four chambers.
4. Exertion increases heart____.
5. carries oxygenated blood away from the heart
6. Hemocoels are found in organisms with____circulatory systems.
7. Your dog's circulatory system is of the____type.
8. Fish pump blood through the____to pick up oxygen.
10. Some____cytes are phagocytic.
12. ____veins carry oxygen-rich blood.
13. Blood = cells +_____
14. auricle or a_____
15. sinus____
17. A snake has a____ee chambered heart
19. atria and ventricles
20. aortic____lunar valve
22. cone-shaped region of the heart, the____icle
23. Capi____es connect veins and arteries
25. Chemical studied in lab for its effect on heart rate (first five letters backward)
26. My cat's heart has____ur chambers.

(See page 208 for answers.)

LABORATORY 9
WATER BALANCE AND EXCRETION OF WASTE PRODUCTS

OBJECTIVES: 1. To observe how plants and animals obtain and conserve water.
2. To observe how plants and animals rid themselves of waste products.

WATER BALANCE IN PLANTS

Water is utilized by plants in photosynthesis and in support (turgor). As a solvent it is also essential in transport of minerals (from the ground to various parts of the plant) and photosynthetic products (from the leaf to the remainder of the plant).

Many plant activities are concerned with obtaining and conserving water. There are numerous openings **(stomata)** in the epidermis of leaves. The stomatal pore is surrounded by specialized epidermal cells **(guard cells)** which regulate the pores. When open, air diffuses into the leaf and the carbon dioxide of the air is used in photosynthesis. However, during the time that pores are open, water vapor diffuses through the pores into the drier outside air. This loss of water vapor is known as **transpiration.** A waxy cuticle covering the outer epidermal cell wall prevents water loss through cell membranes.

To replace water lost from stoma and used in photosynthesis, the plant body must supply water in the same quantity as is lost — that is, a water balance must be maintained. Root hairs are primarily responsible for replacement of water — absorbing it from the soil. Once absorbed, xylem cells then transport the water through the plant body. (Refer to **Figure 6.1** to review plant stem structure.)

Transpiration in Plants (Work in groups of four)

1. Half of the tables will set up experimental flasks, and the other half will set up controls.
2. Fill a beaker with water and set up the apparatus as shown in **Figure 9.1.**
3. Hold the stem below water and make a fresh cut across its base. Insert the freshly cut section of *Nandina* through the cardboard so that it extends into the water.
4. Seal the hole around the stem with petroleum jelly. (It is important that this opening is thoroughly sealed.)
5. Place the apparatus approximately 12 to 15 inches from a light source.
6. If your table has been selected to set up a control, the apparatus is prepared as directed in 1-4 above, with the exception that both surfaces of the leaves should be coated with petroleum jelly.
7. Observe the flasks at intervals throughout the laboratory period. Look for small droplets of water on the interior surface of the flask. Save the cardboard. Discard twigs coated with petroleum jelly.
8. **Results:**

   Beaker containing petroleum jelly-coated leaves (control flask):

   Experimental flask:

   Did you observe any relationship between location of stomata on leaves and location of droplets in the flask? (Wait until completion of the next two parts of this exercise before answering this question.)

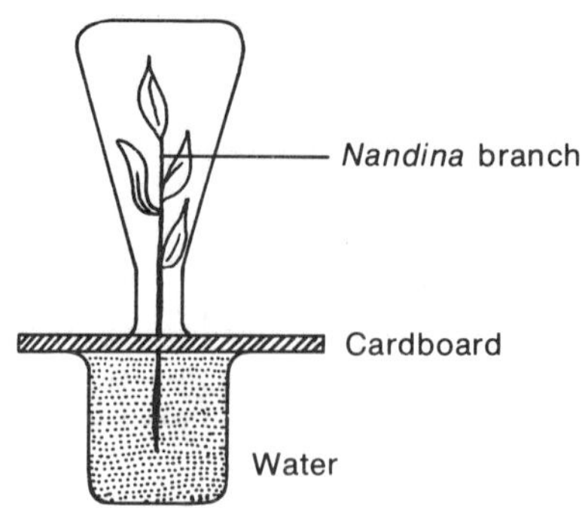

Figure 9.1. Transpiration apparatus.

Cross Section of a Leaf

1. Observe the prepared slide of a dicot leaf.
2. Beginning at the upper surface of the leaf and moving to the lower surface, the following structures may be noted.
   cuticle — waxy outer covering.
   epidermis — single thin layer of cells in this plant.
   palisade cells — in this plant a single layer of elongated cells containing many chloroplasts. Principal site of photosynthesis.
   spongy layer — irregular cells immediately below palisade layer. Many air spaces separate these cells.
   epidermis — single layer with frequent stomata, each with a pair of surrounding guard cells.
   cuticle
3. Sketch a representative section.

4. Conclusions:
   Location of stomata confined to ______________________________

   Possible reason:

   Location of palisade cells confined to ______________________________

   Possible reason:

Location of air spaces confined to ______________________________

Possible reason:

Number and Location of Stomata

1. Cut midrib from a *Nandina* leaf.
2. Place half of the leaf **(upper surface down)** on one end of a glass microscope slide.
3. Place one drop of acetone on the leaf and immediately place a **plastic** coverslip down on the leaf.
4. Press down firmly on the coverslip for approximately one minute. (Be certain you do not have acetone on your finger.)
5. Remove leaf and discard. Repeat the process using the top surface of the other leaf half.
6. Coverslips may then be placed on a drop of methylene blue (with etched side down) and observed for location and relative numbers of stomata.

Leaf Surface	Stomata Present	Relative Numbers
Upper	______________	______________
Lower	______________	______________

QUESTIONS:

1. What effect should fanning the leaf have on the rate of transpiration?

2. What effect should increase in light intensity have on rate of transpiration? Explain.

3. Did water droplets occur in the experimental flask in the transpiration experiment? Explain.

4. Did the control transpiration flask develop water droplets? Explain.

WATER BALANCE AND WASTE EXCRETION IN ANIMALS

Water is essential to life, acting as a general solvent in the body and as a medium for metabolic processes. Body fluids and cell cytoplasm contain a number of dissolved ions such as sodium, potassium, magnesium, chloride, bicarbonate and phosphate. A change in concentration of these ions can cause alterations in many body processes. For example, increase in potassium affects heart rate and changes in magnesium affect nerve function.

Animals living on land, in the sea and in fresh water have solved in different ways the problems of maintaining a constant internal environment. Marine invertebrates generally have an osmotic concentration in their body fluids which equals that of the surrounding sea water. Marine bony fishes have an internal concentration less than that of the sea water and so must constantly take in water to avoid dehydration. Body fluids of fresh water invertebrates and fish, as well as amphibians, are hyperosmotic to the surrounding water, but lose excess water by forming urine of very low osmotic concentration. Terrestrial animals must take in water to equal the amount lost, if they are to remain in balance. Water is lost through evaporation from lungs and skin, excreted in urine, and eliminated in feces. It is gained from oxidation of foodstuffs, from ingested foods, and from drinking water.

Excretory organs function in conserving as well as in excreting excess water. They also play a major role in the elimination of many dissolved substances. The principal excretory problem that we will study is the elimination of nitrogenous wastes from protein breakdown. Common end products of protein degradation are **ammonia** (NH_3), **urea** ($CO(NH_2)_2$) or **uric acid.** Ammonia is highly soluble in water but is toxic. In many aquatic animals this is the end product of protein breakdown, but since it is excreted into the surrounding water, concentrations do not reach toxic levels. Terrestrial animals convert ammonia to more complex, but less toxic compounds prior to excretion. Urea, produced by amphibians and mammals, is soluble in water and has a much lower toxicity. It is secreted in a liquid urine and thus results in water loss. Most birds, reptiles, insects and land snails produce uric acid as a final excretory product. This compound is almost insoluble, thus allowing water to be reabsorbed and the uric acid excreted in a semi-solid urine with very little water loss.

End Products of Protein Breakdown

Observe molecular models of urea, uric acid and ammonia.

Record the number of carbon atoms (black) and nitrogen atoms (orange).

Urea		Uric Acid		Ammonia	
C	N	C	N	C	N

Which compound is most efficient in getting rid of nitrogens? ______________

Which compound is most wasteful in loss of carbons? ______________

Tests for Various Nitrogen Containing Compounds
Setting Up Positive and Negative Controls

CAUTION: Nessler's solution contains sodium hydroxide. Wash it off immediately if you get it on your hands. Use droppers in containers with snails and *Paramecia.* **Do not** use one of your droppers in these containers.

1. Mark one dropper at each table so that it will be used **only** with Nessler's solution. It is important that this dropper not be used for the test with urease, as mercury in the Nessler's is an inhibitor (inactivator) of the enzyme.
2. Set a slide on a piece of white paper. Use a grease pencil to mark a line between the two tests.

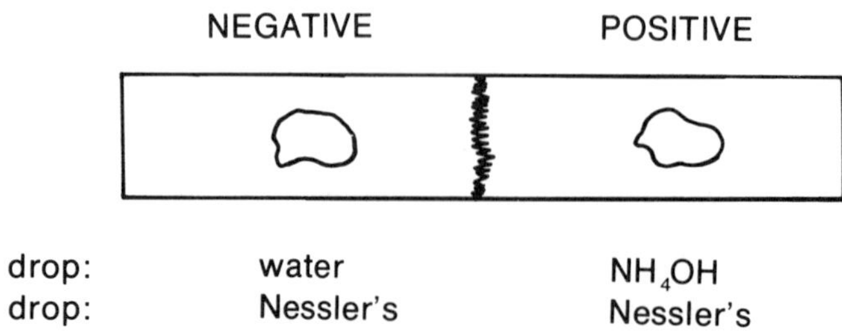

	NEGATIVE	POSITIVE
1 drop:	water	NH_4OH
1 drop:	Nessler's	Nessler's

3. Add materials as shown. Describe the color and appearance of each control.
4. Save this slide and the next one so that you can compare them with your unknown.

Reaction of Urea with Nessler's Reagent

Urea, another end product excreted by some animals, is formed from ammonia. Breakdown of urea by the enzyme urease releases ammonia. The essential features of this reaction are:

$$\text{urea} \xrightarrow{\text{urease}} \text{carbon dioxide} + \text{ammonia.}$$

1. Place approximately 1 ml (1/4 inch) of 2% urea solution in a test tube.
2. Add a few grains of urease (pick up some of the powder on the end of a toothpick).
3. Mix thoroughly and let stand 1-2 minutes.
4. Set a slide on a piece of white paper and add materials as shown. Describe the color and appearance of each test.

1 drop:	urea	(urea + urease) mixture
1 drop:	Nessler's	Nessler's

Which of the four tests just performed are controls?

Excretion of Nitrogenous Products by Living Organisms

A number of freshwater snails have been placed in a small quantity of water and left overnight. Waste products excreted by these animals should be relatively concentrated, since the quantity of water is small.

1. Test a drop of water from the snail container using Nessler's reagent.

(CAUTION: Use only the dropper in the snail container to avoid contamination and possible erroneous results.)

2. Test Results:

3. Conclusion: End product excreted by snails is ______________________.

Effect of Nitrogenous Products on Living Organisms

Urea is rapidly broken down to ammonia by bacterial action. As this presents problems in testing for urea as an end-product, no test will be conducted on excretory products consisting principally of urea. To show that urea and ammonia may be toxic to life if present in high enough concentration, perform the following tests.

1. Stir the container of pond water and place a drop on a slide. **USE ONLY THE DROPPER IN THAT CONTAINER.** Observe microscopically the various types of organisms present.

2. Place a drop of 2% urea at one edge of the coverslip. Apply a piece of paper towel to the opposite edge. This will pull the urea solution under the coverslip. Repeat this procedure three times to be certain that the organisms are suspended uniformly in the urea solution.

3. Observe the various organisms. Note reactions. Continue to observe 1-2 minutes and record observations.

4. Results: (with 2% urea)

5. Repeat steps 1-3 using a fresh drop of pond water and 0.1% NH_4OH solution.

6. Results: (with ammonia)

7. Which end product is more toxic?

Vertebrate Kidney

1. Observe demonstration slide of vertebrate kidney tissue. Compare it to **Figure 9.2.** How much of this figure is apparent on the slide? Add labels to show **renal corpuscle, Bowman's capsule, tubules.** (Use text to help you.)

2. Compare **Figure 9.3** to the preserved vertebrate kidney, which has been injected with dyes to facilitate identification of structures. Be able to identify **cortex, medulla, ureter, pelvis** and blood vessels.

QUESTIONS

1. In general, would you expect a terrestrial organism to excrete a greater volume of urine in the summer than in the winter? Explain. Assume that the same amount of water is consumed all year around.

2. Is the excretion of a particular end product related to the habitat of the animal? __________ Give examples to support your answer.

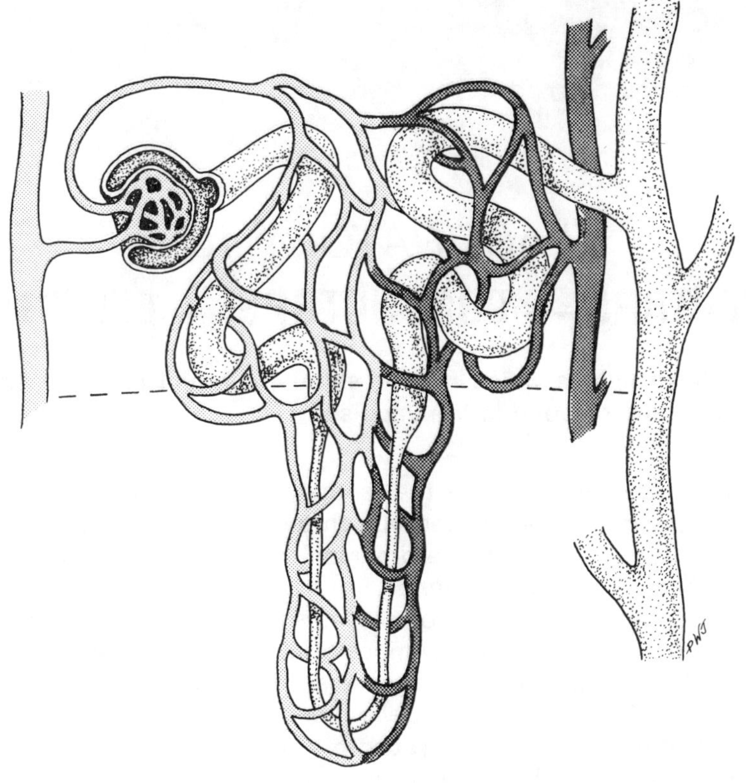

Figure 9.2. Diagram of urine-forming structures in the vertebrate kidney.

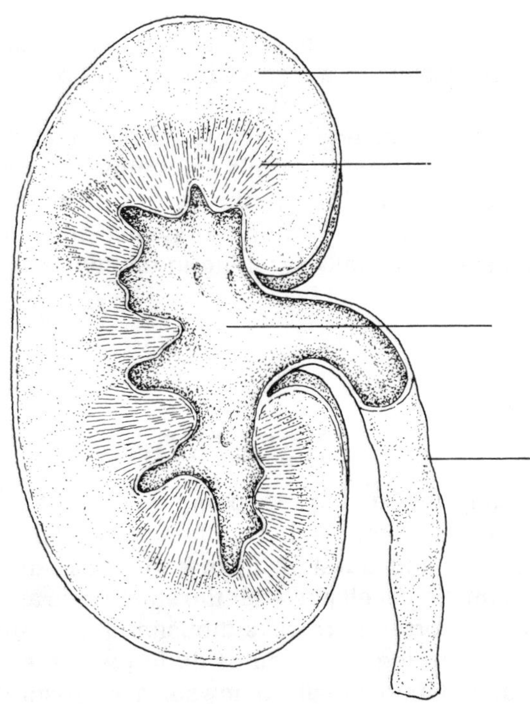

Figure 9.3. Longitudinal section of a vertebrate kidney.

LABORATORY 10
RECEPTORS — CONDUCTORS — EFFECTORS

OBJECTIVES: To study structures which receive stimuli, conduct impulses to other parts of the organism, and respond to the impulse.

A basic characteristic of life is **irritability** — the capacity of the organism to respond to stimuli from its environment. This property can be divided into three phases: (1) a stimulus, or environmental change, causing (2) conduction through the organism of some signal triggered by the stimulus, eventually producing (3) a response, or change, in the organism. It is the organism's gallery of responses that makes up its behavior. All protoplasm is sensitive to at least some types of stimuli, including differences in pressure, chemical composition, temperature of the environment and certain types of radiation. Many organisms possess specialized organs, cells or cell organelles involved in one or more of the areas of responsiveness. Cells specialized for reception of some particular type of environmental information are known as **receptors. Conductors** make up part of the nervous system. The role of the nervous system in the behavior of animals is one of conduction of a nervous impulse from a sensory receptor, modifying or changing it in some way and then channeling information to the appropriate effector. The response itself is performed by **effectors,** which are usually involved in some way with movement of the organism. It should be noted that, in general, the only way to measure the effectiveness of a stimulus is by the nature of the response to it. The impulse pathway determines the ultimate response. In some simple organisms, this pathway leads directly from a receptor to an effector, perhaps involving only a single nerve cell. The more complex the organism's range of behavior, the larger the number of nerve cells which become involved, and the more complicated the patterns in which they are arranged. This complexity seems to have reached its highest level in man, and results in a tremendous range of behavior patterns largely under his conscious control.

Complex, specialized sense organs and receptors are much more characteristic of animals than of plants. Plants are responsive to many different environmental conditions, but their reactions are usually somewhat slower, and occur through the tissues of the plant involved with its growth and maintenance. Only in animals do we find a nervous system specifically concerned with information transfer, and a behavioral system usually resulting in a change in location of the entire organism.

RECEPTORS

While receptors vary greatly, they do possess common properties. A response is elicited only when the **threshold level** of stimulus is reached. This threshold depends on the species of organisms involved, the type, frequency, and duration of the stimulus, and the condition of the individual organism. Some period of time must elapse after the stimulus is applied before the response begins. This time is frequently quite short — perhaps a thousandth of a second. Also, there is some necessary short recovery period following the stimulus during which no new response can be produced regardless of the level of stimulation. Finally, many surface sensory receptors exhibit the property of adaption or **fatigue.** A change in the level of some environmental factor acts as a stimulus, but the receptor gradually becomes adapted to this new level of stimulation and ceases to respond to it. A further response is evoked only by a significantly different level of stimulation.

Sensory receptors are somewhat difficult to categorize. One system groups receptors by the kind of stimulus to which they are sensitive.

1. chemical (taste and smell)
2. mechanical (touch, pressure, hearing and balance)
3. thermal (heat and cold)
4. photic (light and/or vision)

We will now study some different receptors. For these exercises work with your lab partner. If time permits, take turns being the "subject" and the "examiner" on each experiment. If time is short, alternate roles with only one person being "subject" for a given test.

Chemical Receptors

Animals are sensitive to a variety of chemicals present in their environment. Receptors responding to these substances may be distributed over all moist body surfaces or restricted to specific regions. The sense of taste is mediated by receptors clumped together in taste buds on the surface of the tongue. There are several receptor types, each active in sensing a specific primary taste such as bitter, salty, etc. Distribution of these receptors is uneven, but can be mapped, as we will see.

Distribution of taste receptors

1. Designate one student to obtain small beakers containing a **very small** quantity of the following solutions for your lab table **(label beakers!). Do not** fill beakers.

10% salt
2% acetic acid (sour)
0.1% quinine sulfate (bitter)
5% sucrose (sweet)

2. A student will determine distribution of receptors for his partner and vice versa. One student will close his eyes and gently pinch his nostrils shut while the other student of the pair touches the front of the subject's tongue with a cotton applicator dipped in one of the test solutions. DO NOT CLOSE THE MOUTH. The student should attempt to identify what he tastes, if anything. A correct response should be entered as a + in the appropriate blank on the table provided. This procedure should be repeated until each individual has attempted to taste each of the four solutions at the front, sides, and back of his tongue. (See **Figure 10.1.)** The stimuli should be presented at random, and the individual must correctly identify the solution for the response to be recorded as positive (+). It is very important to keep the eyes and nostrils shut and the mouth open while attempting to identify the taste.

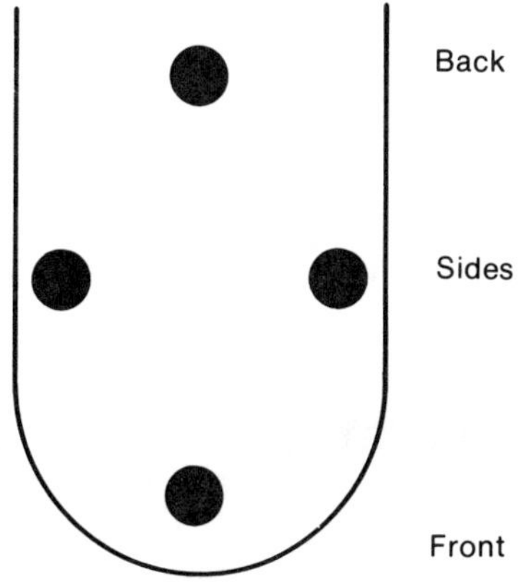

Figure 10.1. Location of areas on the tongue to be used in testing.

3. Record your data in the space provided and on the chalkboard. Each student should also copy the data for the class in the second table for reference in making diagrams of taste receptor distribution.

INDIVIDUAL DATA

	salty	sweet	bitter	sour
front of tongue	______	______	______	______
sides of tongue	______	______	______	______
back of tongue	______	______	______	______

CLASS DATA

	salty	sweet	bitter	sour
front of tongue	______	______	______	______
sides of tongue	______	______	______	______
back of tongue	______	______	______	______

Record the class data as number of correct responses by the group in each area.

Diagrams of your taste receptor areas:

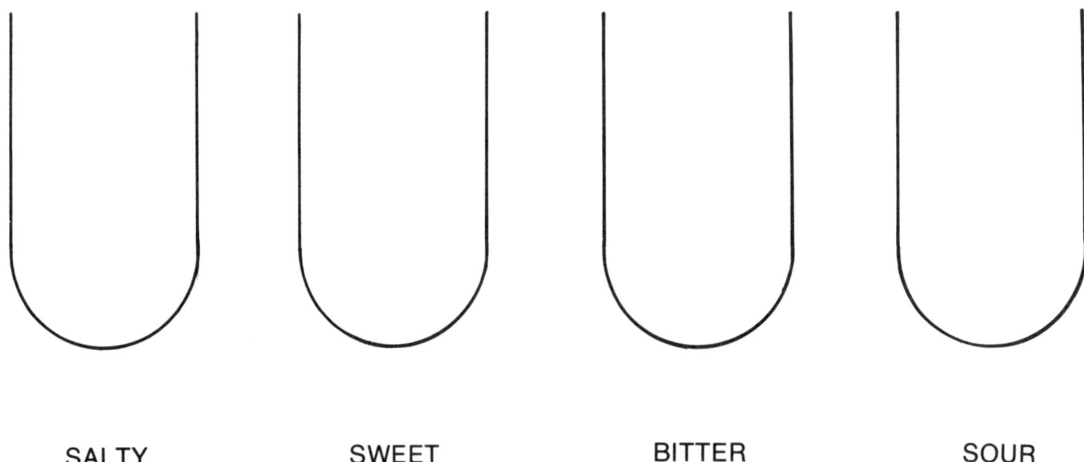

SALTY SWEET BITTER SOUR

Structure of a taste bud (demonstration) (example of a receptor)

Observe the demonstration slide showing the cross-section of the rabbit's tongue. Note the bumps or **papillae** on the upper surface. Lining the papillae are small oval groups of cells — the **taste buds.** Each taste bud has a **pore** leading into the space or **crypt** between the papillae. Each sensory cell has a small hair-like projection pointing toward the center of the taste bud. Diagram one crypt and the walls of the papillae on either side of it.

Mechanical Receptors

1. The pattern of sensitivity for touch discrimination on an area of skin can be determined by measuring the **"two-point" threshold,** the smallest distance at which two stimuli are perceived as two. This measures the density of receptors, since for two stimuli to be felt as two, they must excite two touch spots having at least one unexcited receptor between them.

2. With the points of scissors held the specified distance apart, touch a given skin area of the subject (who must close his eyes). Use either one of the scissors points or both simultaneously. The subject responds by saying "one" or "two" depending on how many distinct touch sensations he feels. Test each spot on the skin three or four times with one or two points of the scissors, but IN RANDOM ORDER unknown to the subject. Nor is the subject told in what order various distances between scissors points are being tested.

3. In the table record a minus sign whenever two points are felt as one, and a plus sign whenever two points are felt as two.

Skin areas	Distance between scissors points			
	3 mm.	6 mm.	1.2 cm.	2.5 cm.
Fingertip				
Palm of hand				
Back of hand				
Outer forearm near elbow				

Which skin area has the highest density of receptors? ______________________

The lowest? ______________________

Thermal Receptors

1. Using a ball point pen, rule off a square 1 cm on a side on the back of the subject's hand, then have the subject close his or her eyes. Straighten the first bend in a paper clip, leaving the other end folded as a handle. Use this as a probe. Place the metal probe in a mixture of ice and water for several seconds, blot off quickly on a paper towel, then gently apply to the ruled area. Repeat. Locate the receptors where **cold** (as distinct from mere pressure) is felt. Use the ball point pen to mark the location of these receptors in the square.

2. Repeat the procedure using the probe warmed in hot water (38-43°C). Mark the receptors which detect the stimulus of heat with a different color pen.

No. of cold receptors/cm² ____________

No. of heat receptors/cm² ____________

Are heat and cold receptors located at the same sites?____________

Photic Receptors

Although the eye has often been compared to a camera due to its lenses and light-sensitive areas, it is physiologically quite different. Within the eye, two cell types are important in vision: the **rods** in the perception of light as such, and the **cones,** used for both color vision and detailed work in optimum light intensity. The spot on the retina where visual images are focused consists almost entirely of cone cells, while rods occupy the peripheral area.

Blind spot — At the point where the optic nerve joins the retina, there is a small region with no rods or cones — your blind spot. Place your hand over your right eye and focus your left eye on the black dot below. Hold the page close to your face and slowly move it away until you find that the black X has disappeared.

What has happened to cause the X to disappear?

How far away from your eye was the book when the X disappeared?____________

Was this distance the same for everyone at your table?____________

CONDUCTORS

Single nerve cells **(neurons)** transmit stimuli from one part of an organism to another. In simple pathways it may be receptor — conductor — effector. More complex pathways are formed by interspersing additional conductor cells between receptor and effector cells. Each neuron contains a nucleus and varying numbers of thin fibers which extend from the irregular cell body. Neurons vary in size, shape and general organization, but some essential features are shown in the representative cell diagrammed in **Figure 10.2.**

Axons vary in number and size from cell to cell. They may extend over relatively great distances, and function in conducting impulses away from the cell body. **Dendrites** receive impulses from other cells and conduct these impulses toward the cell body. Dendrites are generally short, frequently branched, and there are many per neuron. As a general rule, axons are longer than dendrites, may or may not branch, and are usually limited to one (rarely two) per neuron. Concentrations of cell bodies and dendrites are known as **ganglia.** The primary or main concentrations of ganglia make up the central nervous system of the organism. The **central nervous system** consists of the brain and the spinal cord and functions as a receiving station for impulses from sensory nerves and as a transmitting station for impulses to various parts of the body. Those nerves under the control of animals stimulate striated muscles and are called **motor nerves** since motion is usually involved. Nerves regulating internal organs compose the **autonomic nervous system,** generally considered as not under voluntary control of the animal, although some workers disagree with this idea.

Between neurons are found **synapses,** spaces across which impulses are transmitted between the axon of one neuron and the dendrites of the next. In the vertebrate central nervous system, **white matter** is made up of large concentrations of axons and their sheaths. Cell bodies and dendrites appear darker **(gray matter).** Estimates place the approximate number of nerve cells per man at ten billion.

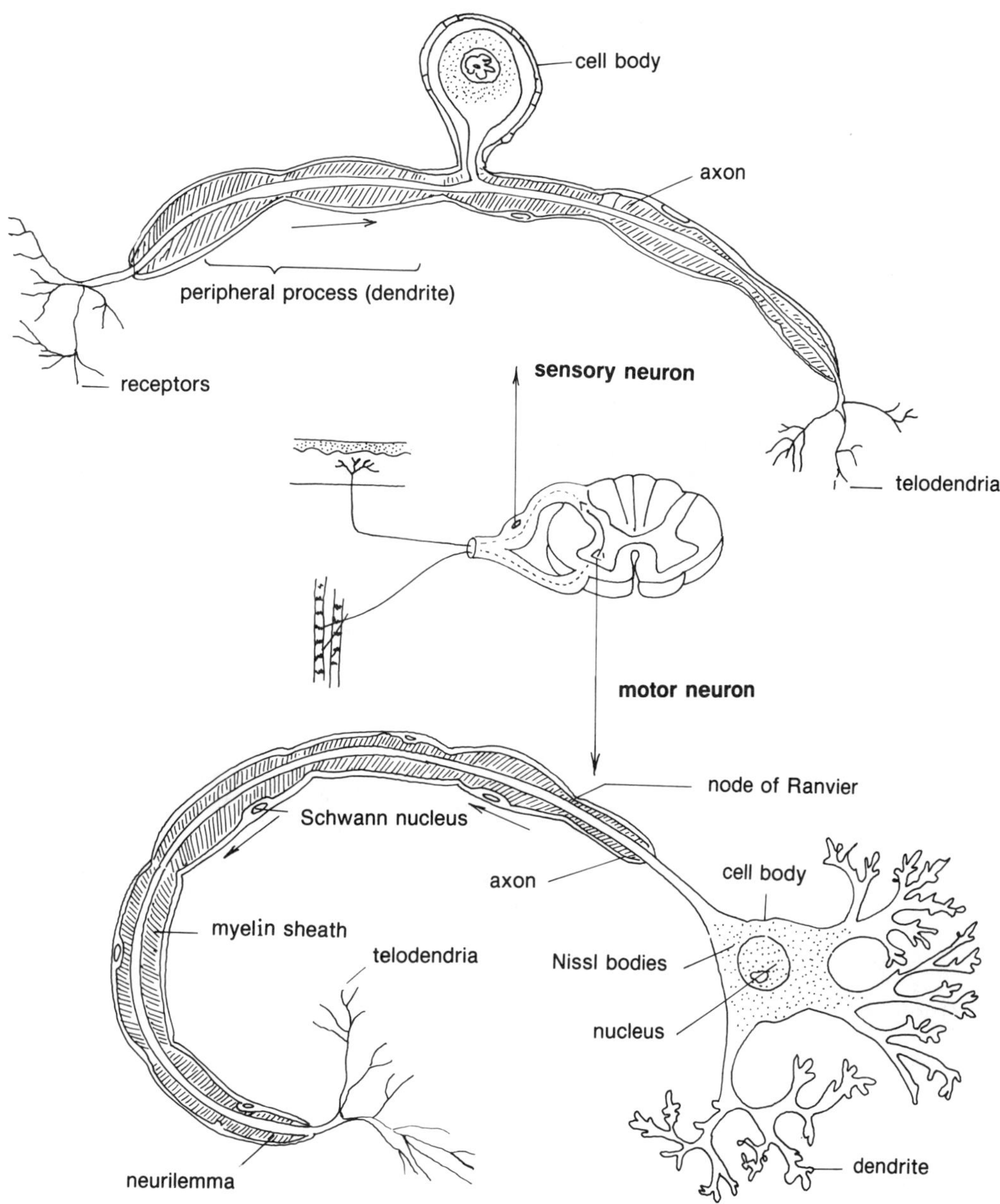

Figure 10.2 Motor and sensory neurons.

Neurons (Demonstration)

Observe the prepared slide of neurons and sketch and label a single nerve cell.

Primitive Nervous Systems

All animals above the level of the sponges exhibit some form of nervous system, although often quite primitive. A **peripheral nerve net** composed of a network of nerve cells is located in or near the surface layer of most invertebrates. Animals such as *Hydra,* jellyfish and anemones do not contain a true nervous system; however, the simplest possible type of true nerve pathway is found in the tentacles of some. It contains the fundamental elements of which all higher nervous systems are built. This simple pathway consists of a receptor-conductor cell and an effector cell. Such a simple system does not allow complex behavior patterns because (a) alternative impulse pathways are not available and (b) no interconnections are made between this pathway and any other part of the nervous system. **Figure 10.3** shows the two-dimensional nerve net of a hydra.

Slightly greater organization is shown by jellyfish and sea anemones. In these animals, the net is concentrated in specific areas to produce rings and bands similar to nerves **(Figure 10.3).** Connections occur between the network and sensory cells in the epithelium and muscle fibers. Impulses may flow either direction along an axon, but there is a greater tendency for them to occur in one direction only. Another advancement seen in this type of system is that different muscles respond to different levels of stimulation. No structure resembling a brain is present.

Evolutionary Trends Seen in Higher Animals

Most free-living animals move in one particular direction so that the same part of the animal is always first to contact the changing environment. As a result, higher organisms have developed **bilateral symmetry** (in which there is only one plane which can be passed in a particular direction through the body axis dividing it into equal right and left halves).

Briefly, one may summarize events in organisms which developed bilateral symmetry: (a) formation of nerve cords making possible better centralization, (b) increased complexity of behavior associated with increasing complexity of pathways, (c) specialization of areas for different functions, (d) unidirectional conduction of impulses, (e) increase in sensory organ numbers and types, (f) increasing dominance of the anterior end of the organism **(cephalization).**

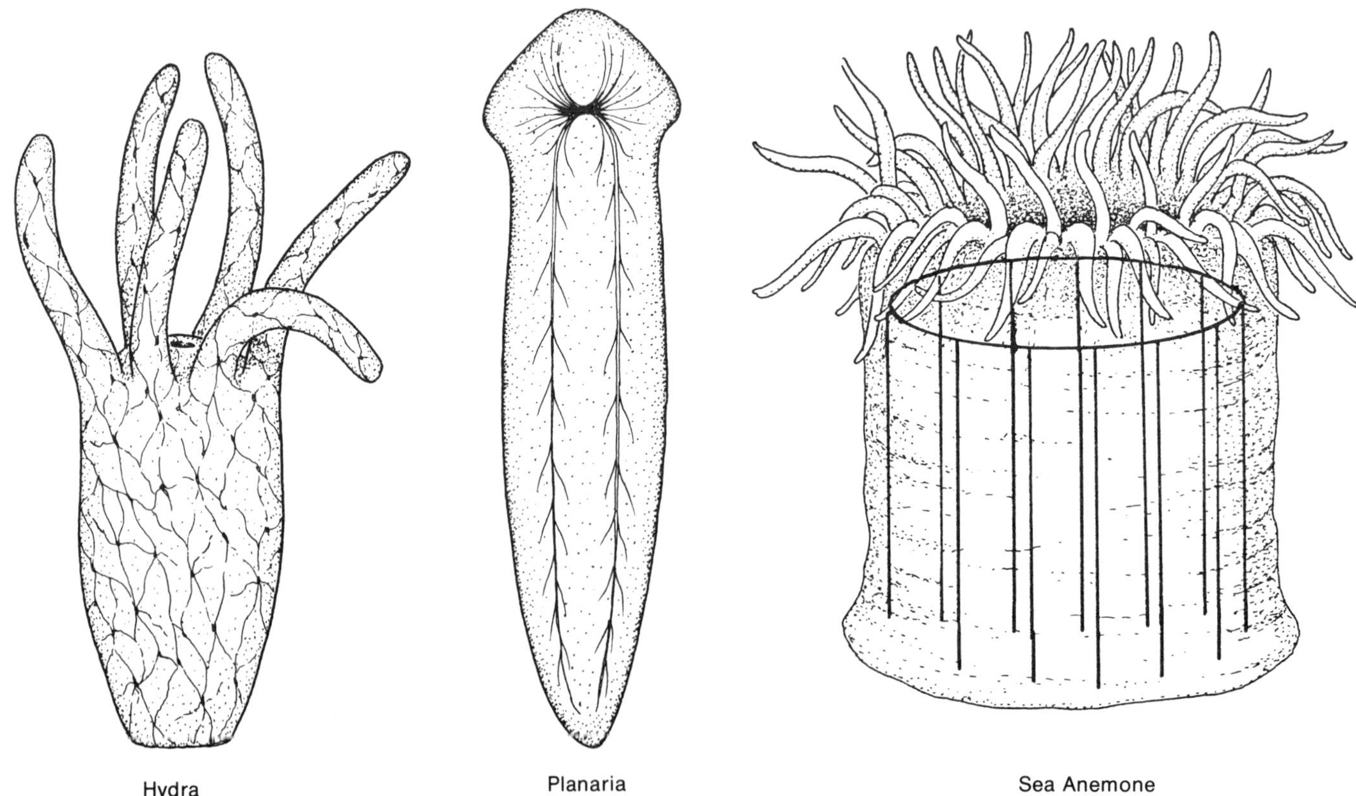

Figure 10.3. Simple nervous systems.

1. Head dominance may be seen to be already well-established in the simple **flatworms** such as *Planaria* **(Figure 10.3).** The pair of swellings at the anterior (front) end of the body are cerebral ganglia or a "brain," the integrative center of the central nervous system. The **three principal components of a nervous system** are seen in the planarian worm: (a) sensory structures — primarily found anteriorly, (b) a brain — also located anteriorly and functioning for relay and coordination of impulses, and (c) unidirectional system of nerve fibers connected to effectors (muscles). In these simple animals one sees the beginning of the evolutionary trend toward increasing dominance of the brain over the remainder of the central nervous system.

2. The **earthworm** is more advanced than the flatworm in many respects. You will study the earthworm next week. In it, the two nerve cords come together to form a double, **ventral** cord with a localized swelling (segmental ganglion) within each segment. Lateral nerves lead from each segmental ganglion to the body wall and parapodia. The brain is located anteriorly and dorsally, thus requiring that nerves from the brain to the ventral nerve cord go around either side of the digestive tract, producing a ring around the esophagus **(Figure 11.2).**

3. In **vertebrate animals** (including fishes, amphibians, reptiles, birds, and mammals), the central nervous system consists of a single, hollow nerve cord located **dorsally** to the digestive tract. Anteriorly the cord is enlarged to produce the brain. A bony or cartilaginous skeleton surrounds the cord, protecting it; the brain is protected by a skull. Segmentation is shown by the spinal nerves which arise from the spinal cord in pairs, and also by the pairs of cranial nerves coming from the brain. A generalized drawing of major regions of the vertebrate brain is shown in **Figure 10.4.**

Telencephalon — Becomes dorsally enlarged in higher vertebrates to form the **cerebral hemispheres** in mammals. Functions as a center for information integration and, in man, for learning. **Olfactory lobes** (concerned with reaction to odors) are located anteriorly and in mammals, especially, enlarge to produce the **cerebral cortex** where most regulatory activity is organized.

Diencephalon — Connects to posterior portions of the brain.

Mesencephalon — Develops into **optic lobes,** which are hidden by the cerebrum in man.

Metencephalon — Grows dorsally to form the **cerebellum.** Functions in muscular coordination, posture and balance. This area is small in amphibians, becoming larger in reptiles and mammals and is very highly developed in birds and fish.

Myelencephalon — Forms the **medulla.** Functions in control of respiration and circulation. Connects the brain to the spinal cord.

4. Compare the generalized drawing of the vertebrate brain to models and dissected specimens available in the lab. Relate the size of the various portions of the brain in the different animals by comparing them to the frog brain and complete the chart.

MODEL OR SPECIMEN	OLFACTORY LOBE	CEREBRUM	OPTIC LOBE	CEREBELLUM	MEDULLA
Reptile (Alligator)					
Bird (Pigeon)					
Fish (Trout)					
Fish (Shark)					
Mammal (Dog or Rabbit)					
Man					

(+ = increased in size relative to frog, − = decreased in size relative to frog, ± = no change)

Compare the relative size of cerebral hemispheres of the reptile to those of the bird. Do the same with the cerebellum. ____________________ have larger (relative to overall size) cerebral hemispheres. ____________________ have larger (relative to overall size) cerebellums. Compare the size of the cerebral hemisphere and optic lobe of the amphibian. Which is larger? ____________________ Compare the reptile cerebral hemisphere and optic lobe size to those of the amphibian. ____________________ have larger optic lobes. ____________________ have larger cerebral hemispheres.

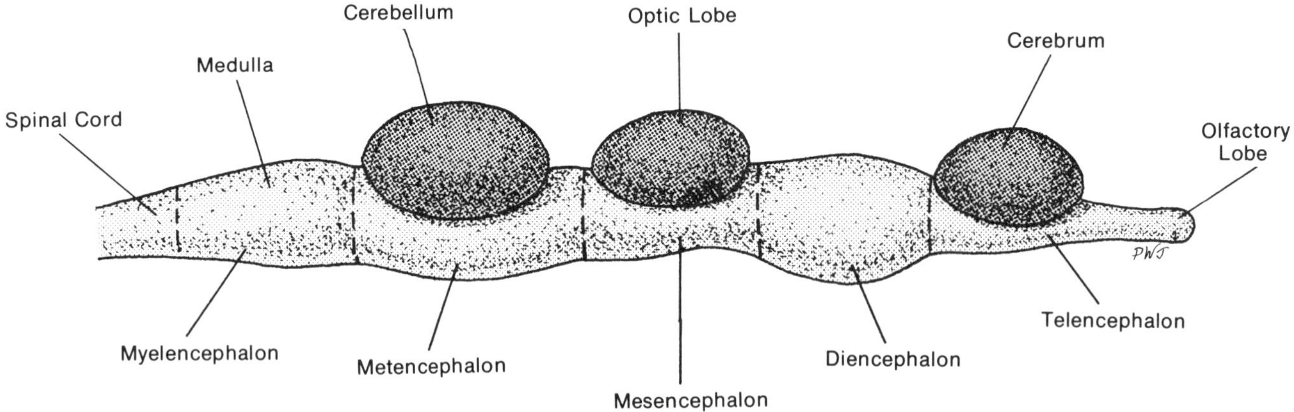

Figure 10.4. Diagram of a generalized vertebrate brain.

QUESTIONS

1. Which animal or model exhibited the largest **relative** size of the following regions:
   a. cerebrum —
   b. olfactory lobe —
   c. optic lobe —
   d. cerebellum —
2. Which animal or model exhibited the smallest **relative** size of the following regions?
   a. cerebrum —
   b. optic lobe —
   c. cerebellum —
3. What relationship exists between nervous systems and body symmetry?

4. What relationship is there between enlargement of certain brain areas and the characteristics of that animal?

EFFECTORS

The final step in the process is the response of the organism to the stimulus. In animals, **effectors** are muscles. The response occurs as a result of the initial stimulus received by the receptors. Muscle tissues were studied in Exercise 6.

LABORATORY 11
ORGAN SYSTEMS IN AN INVERTEBRATE AND A VERTEBRATE

OBJECTIVE: To integrate information obtained in previous labs by observing an animal as a whole unit.

In previous labs you have observed a number of different organisms and studied various isolated organs and organ systems. In this exercise, tissues and organ systems will be studied in an intact organism. As you study each organ system, keep in mind that each depends on all the others, since there is specialization or **division of labor** among cells of a multicellular organism.

ORGAN SYSTEMS OF AN INVERTEBRATE ANIMAL

The earthworm, *Lumbricus terrestris,* is an invertebrate, an animal without a backbone. It is relatively simple when compared with the extremely complex bodies of vertebrate animals. There are similarities, however, in organs and organ systems.

External Anatomy and Beginning Dissection (Students will work in pairs).

1. Obtain a preserved worm. Identify the **dorsal** (top) side by the dark, **dorsal blood vessel.** Note the mouth on the **ventral** (lower) side of the first segment. Arrange the worm so that the dorsal side is uppermost. At about segment 31 is the **clitellum** which is responsible for egg cocoon formation in mature worms.
2. Movement is partly related to pairs of bristle-like **setae** on most segments. Using a hand lens or dissecting microscope, find the openings for the setae on each segment. How many do you find per segment?______Is the number the same on all segments?______
3. Place a moist paper towel in the bottom of a dissecting pan. Extend the worm full length, **ventral surface down,** and pin to the paraffin base at the first and last segments. Add water to the pan to keep tissues moist while you work.
4. With a scalpel, carefully make a small cut through the body wall at segment one and continue cutting **with scissors** down the **anterior** (front) third of the body. **Be careful not to injure underlying tissues.**
5. Pin the body wall back so that internal organs may be seen. As you proceed, carefully separate each **septum** (membrane separating internal segments) from the central tube. The space between each pair of septa is part of the **coelom,** or body cavity. The inner tube is the digestive tract.
6. Around segments 9-13 you will see several off-white, rounded structures over the digestive tract. These are the reproductive organs. The earthworm is **hermaphroditic;** that is, both male and female reproductive organs are located in each individual. Carefully move these light bodies aside to observe the digestive tract.

Digestive Tract

1. Use charts posted in the lab and **Figure 11.2** to help locate **mouth, pharynx, esophagus, crop, gizzard** and **intestine.** The intestinal tube is modified in the last segment to form the **rectum** which opens to the outside through the **anus.** Label **Figure 11.1.**
2. We will return to the digestive tract after observing some of the other organ systems.

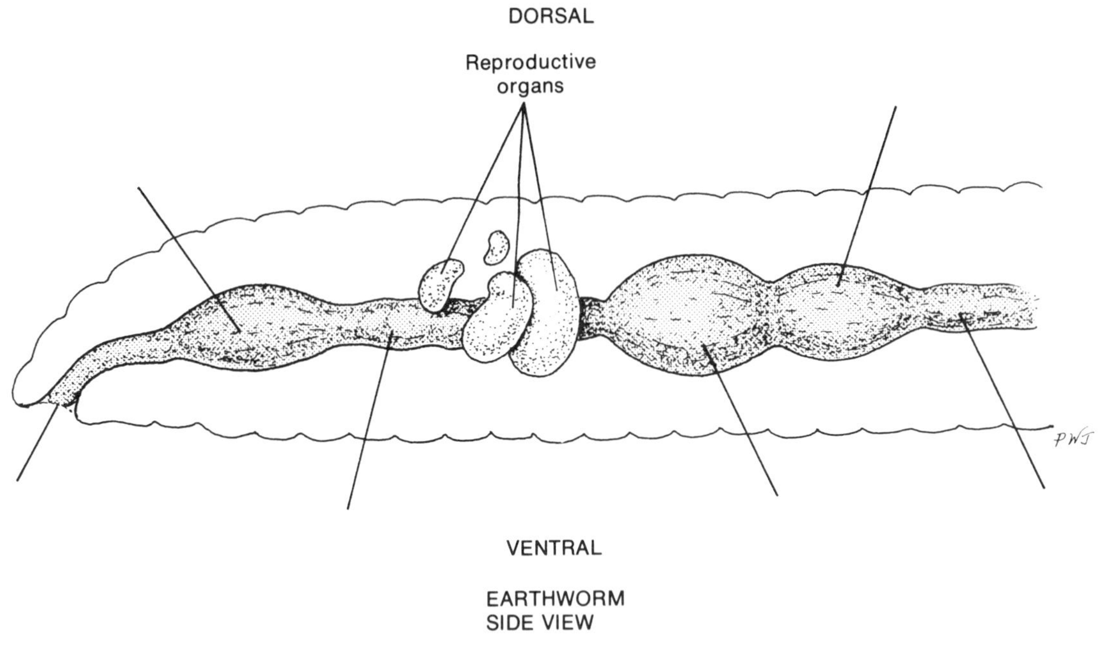

Figure 11.1. Diagrammatic reprepresentation of part of the earthworm digestive system.

Gas Exchange

The earthworm burrows in moist soil. Gas exchange depends on a moist environment and a well developed capillary network close to the skin. Oxygen diffuses across the thin, moist skin to the bloodstream and is carried throughout the body. CO_2 is released in a similar manner (moving in the reverse direction).

Circulation

1. The **dorsal blood vessel (Figure 11.2)** is a collecting tube and pulsates, moving blood from posterior to anterior portions of the body. Locate this vessel, partly imbedded in the top of the intestine.

2. In segments 7-11 there are **five pairs of hearts,** which circle around the digestive tract. These join to the **dorsal blood vessel** which will be seen later when you remove the digestive tract. The **ventral blood vessel** returns blood to the body. Along the length of the worm, capillaries supply internal tissues and organs with blood, and connect the ventral and dorsal blood vessels.

3. Is this an example of an open or a closed circulatory system? ______________________

Excretion

1. These animals have excretory organs known as **nephridia,** which remove wastes from both tissue fluids and blood, and thus have properties similar to kidneys. A pair of nephridia (one on each side) is located in each segment except the first three and the last.

2. In the region of the gizzard and posterior to it, examine the inner surface of each segment. Locate the nephridia, which are seen as small, tan structures under the digestive tract.

3. Review nephridia structure in your text. If you wish, you may carefully remove an entire nephridium from the animal and mount it in a drop of water on a slide. Add a coverslip and observe under the compound microscope. You may be able to identify some of the parts. In addition, look for tiny nematode worms, which often parasitize the earthworm.

Digestive System (continued)

1. With a scalpel and a probe or forceps, carefully remove the digestive tract by making a cut across it in segments 5 and 22 or 23, and gently lifting from underneath.

2. Cut across a portion of the pharynx, esophagus, crop, gizzard, and intestine. Compare the thickness of the muscular walls in each of these areas.
Make a list, arranging them in order of decreasing thickness. _______________ _______________

_______________ _______________ _______________

3. How does the thickness of each organ relate to its specific function in digestion? Refer to your textbook if you need a clue.

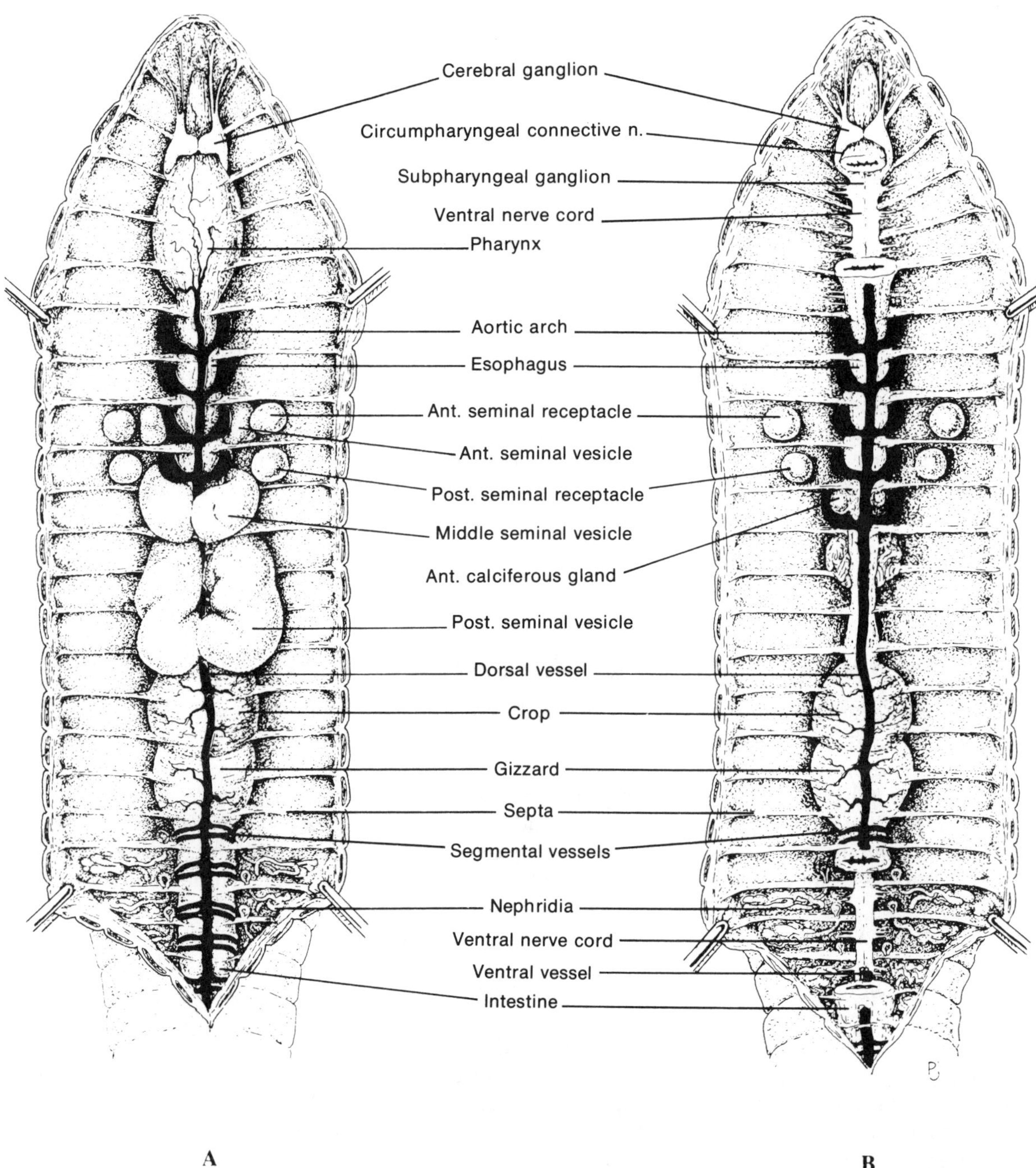

Figure 11.2. Dorsal views of the internal features of *Lumbricus terrestris.* **A,** Nephridia in segments 4-19 removed; **B,** Pharynx, seminal vesicles and a section of intestine removed.

Nervous System

1. The "brain" is located in the region of the third segment on the dorsal surface and is connected to the ventral nerve cord by nerves encircling the digestive tract.

2. The **nerve cord** may be seen as a white thread on the ventral wall. Remove a segment of this cord and observe with a dissecting microscope or hand lens, using a black background. Note its double nature and the presence of **segmental ganglia** (swellings of the nerve cord in each segment), and three pairs of **nerves** on each ganglion (two pairs close together near the posterior portion of the segment and one pair in the anterior portion).

3. With a teasing needle, follow the nerve cord anteriorly and note how it encircles the digestive tract in the region of the "brain." (It may be necessary to remove more of the digestive tract so that you can follow the cord to the anterior end of the worm.)

Effectors

1. Although one may not ordinarily think of the earthworm as having a skeleton, the action of body-wall muscles on the incompressible fluids within the body bring about very effective movement. We describe this arrangement as a **hydrostatic skeleton.**

2. Since the body is divided into compartments, a compartment which contracts in diameter must simultaneously increase in length. Thus, the earthworm's fully segmented body is capable of a great variety of movements. Forward movement is accomplished by a contraction of circular muscles accompanied by relaxation of longitudinal muscles. Following the push forward, there is a contraction of the longitudinal muscles, accompanied by a relaxation of circular ones; this expands the anterior region of the worm against the soil, and the animal is anchored by the setae.

3. Observe a prepared slide of the earthworm cross section and refer to **Figure 7.3.** Just under the epidermis are the two muscle layers. Find the outermost layer (the thinner of the two). This is circular muscle, contraction of which reduces the diameter of the worm and elongation results. The thicker layer is the longitudinal layer. It is continuous from segment to segment.

ORGAN SYSTEMS OF A VERTEBRATE ANIMAL

Now, let's make a similar study of the frog, an amphibian and a member of the vertebrate group. As you locate structures and organs, notice how similar their arrangement is to the organization of man's own. *Homo sapiens* is also a vertebrate animal.

Due to the complexity of internal arrangements, we will identify structures by their positions relative to each other rather than attempting to study each organ system as a completely separate entity. Following the dissection you will be asked to group the organs into the systems of which they are a part. Charts and models are available to help you identify structures. Students will work in pairs.

1. Place a preserved frog ventral side up on a moist paper towel in a dissection pan and pin down the feet. With forceps, lift the skin of the lower abdomen and make a mid-ventral incision from the hind legs to the lower jaw. Make a cut perpendicular to this first one at each end so you can pin back the skin to expose the musculature.

2. Repeat the cutting pattern described above, this time through the muscle layer, being careful not to damage the organs lying beneath. You will have to cut some small bones near the jaw. Pin back the body wall on both sides.

3. If your specimen is a female, the abdomen may be filled with black, gelatinous masses of eggs. Remove these so that organs will not be obscured.

4. Locate the **heart** within the membranous **pericardial sac (Figure 11.4).** How many chambers are in the frog heart?__________ Locate these chambers. Number of atria__________ Number of ventricles__________ Why does the ventricle have a thicker wall than the atria?

5. Around the heart find the three brown lobes of the **liver.** Close to the middle lobe you will see a small greenish organ, the **gall bladder (Figure 11.3).**

6. The **stomach** projects from under the left lobe of the liver. It receives food from the **esophagus,** leading from the mouth. The stomach empties into the **small intestine.** Follow this narrow tube as it coils around the abdominal cavity. Notice the **mesenteries** which hold it in place to the body wall.

These mesenteries are richly supplied with blood vessels. Why?______________________________

____

7. Near the posterior end of the stomach, suspended in the mesentery, is the long, flat, pinkish or yellowish **pancreas.**
Where does its duct empty into the digestive tract?____________________

8. The small intestine joins to the much shorter, thicker **large intestine** near the posterior end of the abdominal cavity. The expanded portion of the large intestine **(cloaca)** collects products from the bladder, intestine, genital ducts, and liver ducts.

9. Look for the very thin-walled, sac-like **bladder** on the ventral side of the cloaca. Where is the urine formed that comes to be stored in the bladder and is later emptied into the cloaca?

____

10. The red, spherical **spleen** may be found near the cloaca, in the mesenteries.

11. Now remove the liver and heart from the frog. Make a cut across the esophagus near the mouth, and across the posterior end of the large intestine. Carefully cut through the abdominal mesenteries to remove the entire digestive tract. Do not destroy other organs.

12. As you did with the earthworm, cut across the stomach, small intestine and large intestine. Compare the thickness of the walls and relate this information to the function of the organs.

Stomach ____

Small Intestine ____

Large Intestine ____

13. In the abdominal cavity you should now be able to see the elongated, reddish-brown **kidneys,** lying on either side of the midline. What is their function? Around the anterior end of each kidney are yellowish, finger-like **fat bodies.** On the surface of the kidney, a small, flat, light-colored line is the **adrenal gland.**

14. Reproductive organs are adjacent to each kidney. In the male, yellow, bean-shaped **testes** are on top of the kidneys. In the female, **ovaries** are found in the same position. You may have already removed the ovaries if they were filled with eggs **(Figure 11.4).**

15. The **lungs** are in the anterior part of the cavity, probably in a collapsed state. Remove one carefully and cut it in half. The inner surface of the lung is spongy, composed of a network of ridges dividing the wall into a large number of small chambers, the **alveoli.** Blood vessels run along the ridges and break into an intricate network of capillaries in the walls of these alveoli. Why would this be of importance to the function of the lungs?______________________________________

16. Note the vertebral column with the pairs of **nerves** extending outward. Follow some of these nerves and note how they extend into the limbs, the abdominal wall, etc. How many pairs of nerves can you count?______________________________

17. Now, remove your dissecting pins and turn the frog dorsal side up. Repin the limbs. Remove the skin from the head.

18. Using scissors, make a cut through the muscles slightly anterior to the attachment of the front left limb. Cut toward the vertebral column. Repeat on the right side. Gradually cut away the tissues covering the brain, taking care to avoid damage to the very fragile nervous tissue, especially as you chip away the thin bone of the skull.

19. When all overlying tissues have been removed, observe the brain noting relative size and location of the various divisions. Add water so as to keep the surface moist. Compare the specimen to the model and to **Figure 11.5.**

20. Make a cut across the most anterior and the most posterior regions of the exposed brain and gently lift it out. Turn it over and observe its ventral aspect **(Figure 11.5).**

21. Now observe the muscle arrangement in the frog, using a hind limb as an example. These muscles are of the **skeletal** type, and are voluntarily controlled. What is the main function of skeletal muscle?____________________

What type of muscle controls the viscera?____________________

the heart?____________________

Are muscles conductors, effectors or receptors?____________________

Remove the skin from an entire hind limb. Notice that individual muscle fibers are 4 to 5 cm long and are bound together by connective tissue to form the larger bundles known as **muscles.** Skeletal muscles are usually attached by one or both ends to bone.

QUESTIONS

1. List the organs you observed which function in each of these systems.

System	Earthworm	Frog
Digestive		
Respiratory		
Endocrine		
Circulatory		
Excretory		
Nervous		
Reproductive		

2. What are the functions of the outer covering of the earthworm? ____________________

____

of the frog? ____

3. What is the relationship between digestive and circulatory systems?

Between respiratory and circulatory systems?

Between excretory and circulatory systems?

4. Compare muscle arrangement in the earthworm and the frog.

Which organism has a more diversified set of movements?

Why?

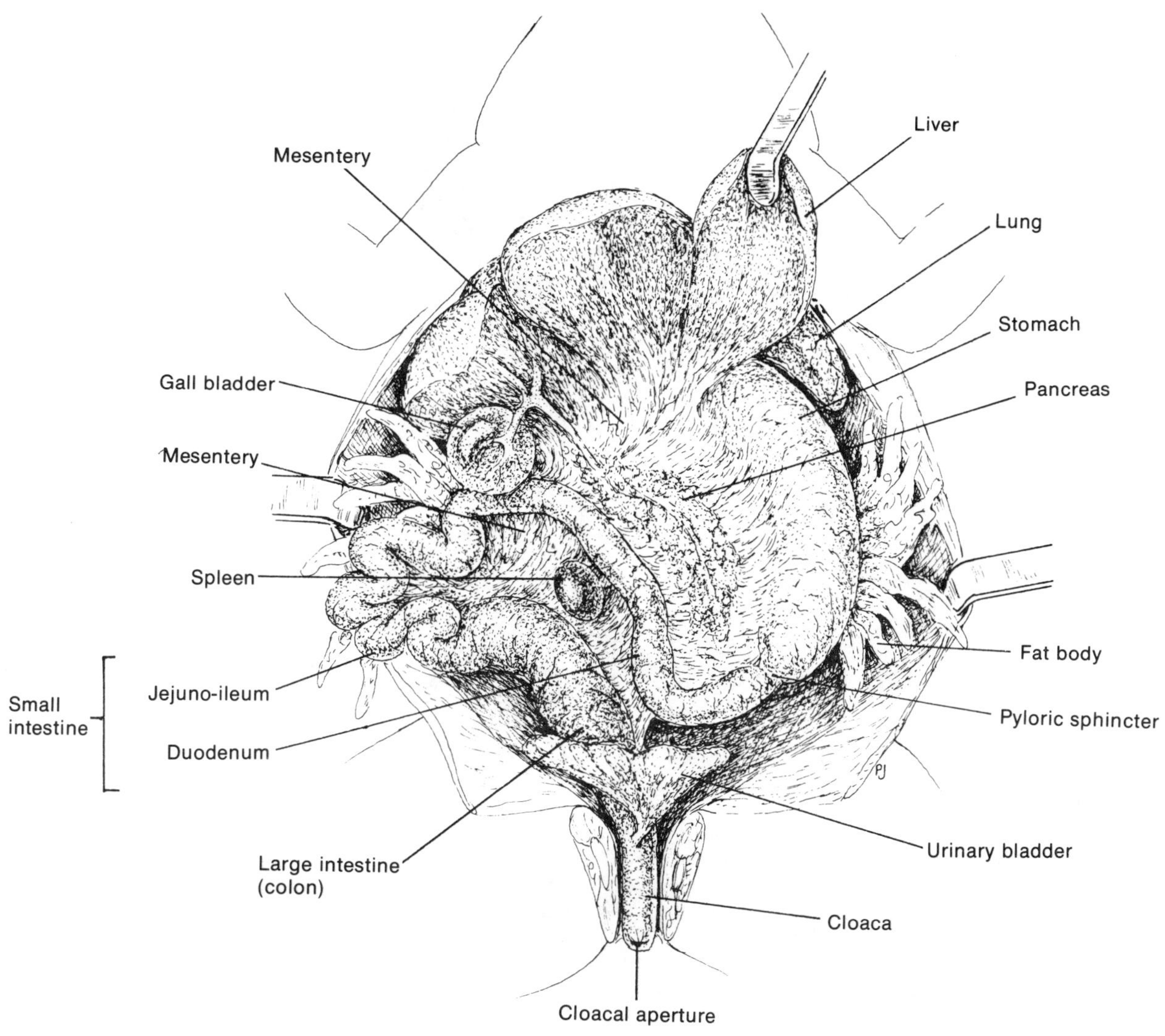

Figure 11.3. The digestive system and other visceral organs of the frog (ventral view).

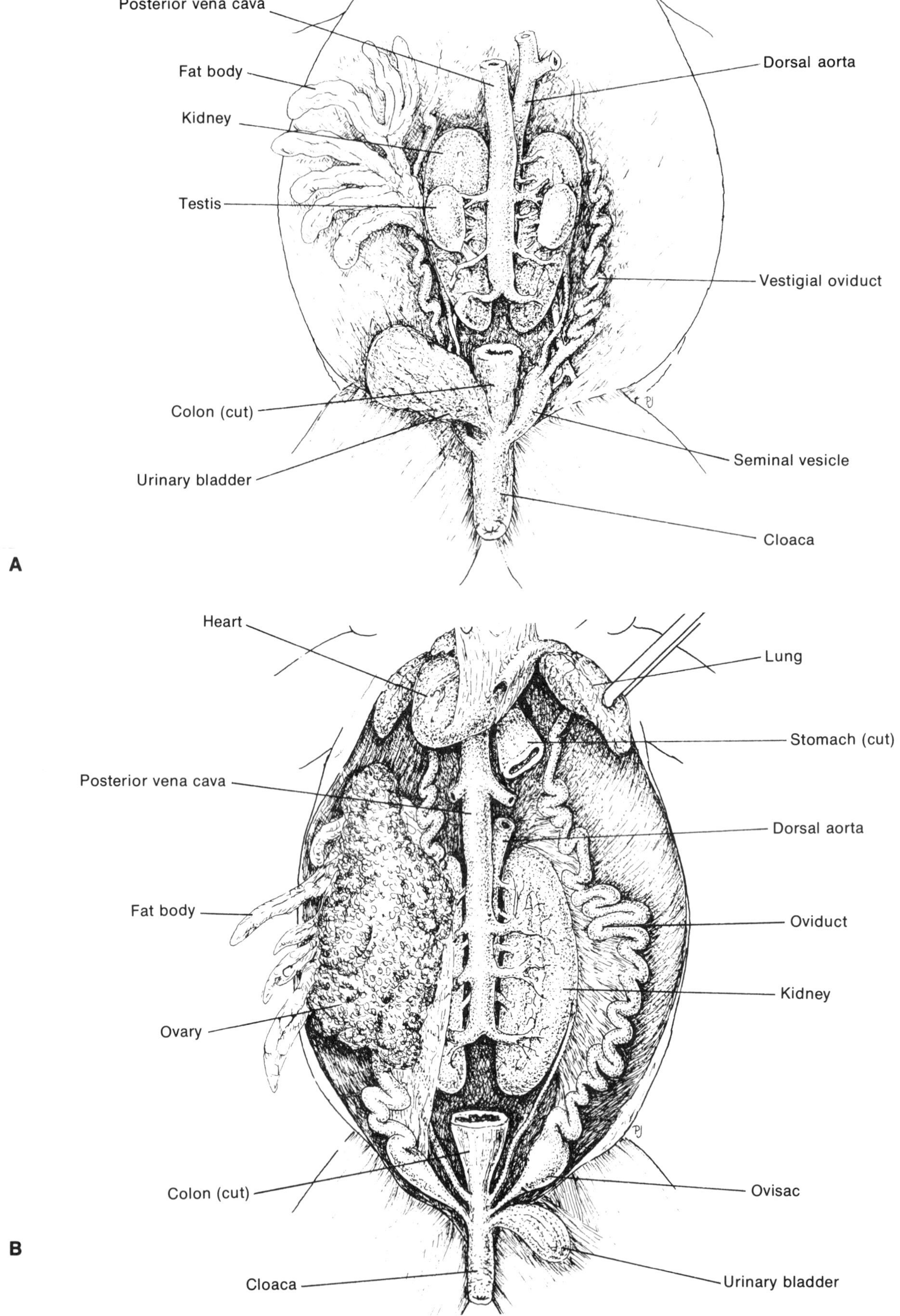

Figure 11.4. A, Male urogenital system of a frog (left fat body removed); **B,** Female urogenital system of a frog (left fat body and the left ovary removed).

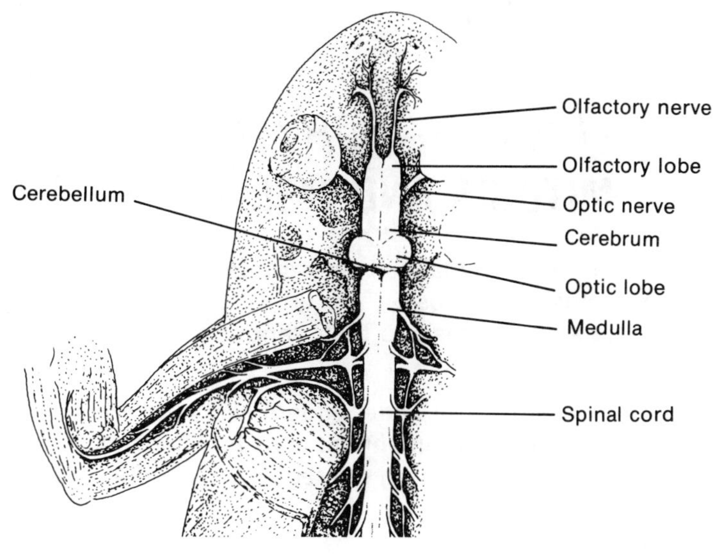

A

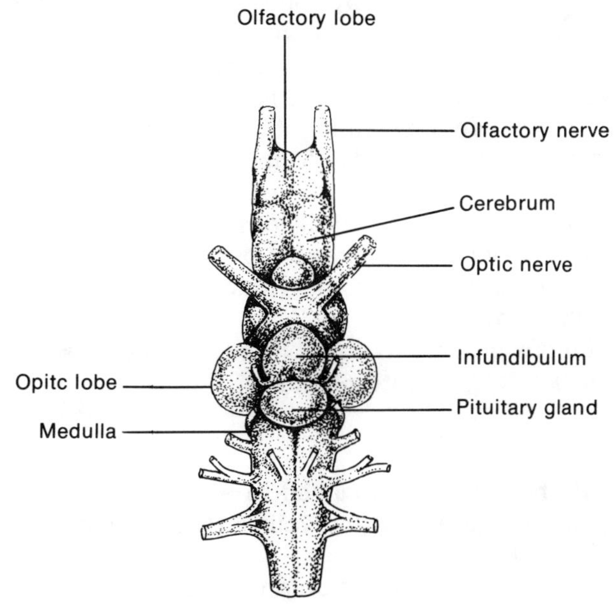

B

Figure 11.5. A, Dorsal view of frog brain; **B,** Ventral view of brain.

LABORATORY 12
MITOSIS AND MEIOSIS

OBJECTIVES: 1. To observe plant and animal cells in various stages of mitosis.
2. To compare mitosis to meiosis.
3. To relate mitosis and meiosis to their various functions in living organisms.

This laboratory is principally designed for independent work in areas where clarification is needed. Questions provide additional reinforcement and may be filled in later if desired, but should time allow, it is a good idea to answer them during the regularly scheduled period so the instructor can assist you in problem areas.

MITOSIS

Mitosis (nuclear division) combined with cytoplasmic division **(cytokinesis)** is the basic form of cell division utilized by eucaryotic cells. A single parent cell divides to form two daughter cells, each genetically identical to the parent (that is, carrying the same hereditary information in the nucleus). This process occurs in growth of multicellular organisms, replacement of damaged or worn-out cells and in asexual reproduction. Formation of two daughter cells is a continuous process with the time required for one mitotic cycle varying from tissue to tissue, from organism to organism, etc. Most cells possess the potential for such division at least once during their life span. Regulatory mechanisms control when and how often a cell divides. Loss of this regulatory mechanism and passage of this "lack of control" from one generation to the next result in cancerous or malignant cells. Such cells divide repeatedly (with no interphase between divisions) and are unable to perform adequately their normal cellular processes.

Although mitosis is a continuous process it has arbitrarily been divided into four different phases. A cell **not** actively dividing is said to be in **interphase.**

Plant Mitosis

Onion *(Allium)* root tips will be stained, squashed to separate the cells, and observed for the various stages of mitosis. Refer to **Figure 12.1** and your text for diagrams and characteristics of the various stages.

1. Obtain 2 whole onion roots or 2 germinated onion seeds from the designated container.
2. With a scalpel, cut 0.5 cm off the **lower tip** of the root. This tip should contain the actively dividing cells. Place the tips on a clean slide supported over a staining tray.
3. Cover the root tips with one or two drops of 9% hydrochloric acid. Allow to stand 10 minutes to soften tissues.
4. Remove the acid with a dropper and add several drops of water to dilute the remaining acid.
5. Remove excess water and cover the tips with a few drops of acetocarmine stain (specifically absorbed by chromosomes). Allow to stand for 10 minutes.
6. Place a clean coverslip over the root tips and **gently** press down on the coverslip with a pencil eraser or a **clean** finger (to spread tissues and cells apart).
7. With low power examine the preparation for areas of small, square cells where active division is occurring. (See **Figure 15.6**).
8. Switch to high power to observe individual cells. On page 105 draw a cell in interphase and in each phase of mitosis.
9. Note the rigid cell walls and the large, darkly stained nucleoli. What is the function of the nucleolus? ______________________

10. Can you count the number of chromosomes in metaphase? ______

11. Note the faint cell plate across the center at telophase. New cell walls will form here.

12. Add the following labels to your drawings: **nucleus, nucleolus, spindle, chromosome, cell plate.**

13. After you have found all of the stages you can locate on your preparation, obtain a prepared slide of the *Allium* root tip and use it to complete your drawings.

Animal Mitosis

1. When you have learned the features of each stage of mitosis, obtain a slide of the whitefish blastula. If you hold the slide over a piece of white paper you can see a series of colored spots, each of which is a section through an early stage in the embryonic development of the whitefish. Cells are dividing rapidly and probably you can find most stages on each slide.

2. Observe with low power. Cells which are essentially uniformly stained are sections which do not include the nucleus. Only a few interphase cells will be noted.

3. Switch to high power and find stages comparable to those seen in the plant cells.

4. You will note certain differences between plant and animal mitotic cells. Note the nucleoli, outer cell boundary, aster and spindle. Also note the differences as the two daughter cells form at telophase.

5. On the same page with the plant drawings, draw any stages which differ from plant mitotic figures.

6. Observe display models of plant and animal mitosis.

The net result of mitosis will be two cells with chromosomes whose hereditary information is identical to that of the original nucleus.

Mitosis Questions

1. List two ways in which mitosis differs in plant and animal cells.

2. The end result of mitosis is two "new cells." Are they really new? Explain what is new and what is not.

3. What is a chromatid and what happens to one during mitosis?

4. What happens to the chromosomes in a cell when they cannot be visualized as distinct chromosomes?

Note: If you do not understand fully the mitotic process **do not** go on to meiosis. Ask for help from the lab instructor and/or refer to the text for clarification. You may want to study the film loop on mitosis.

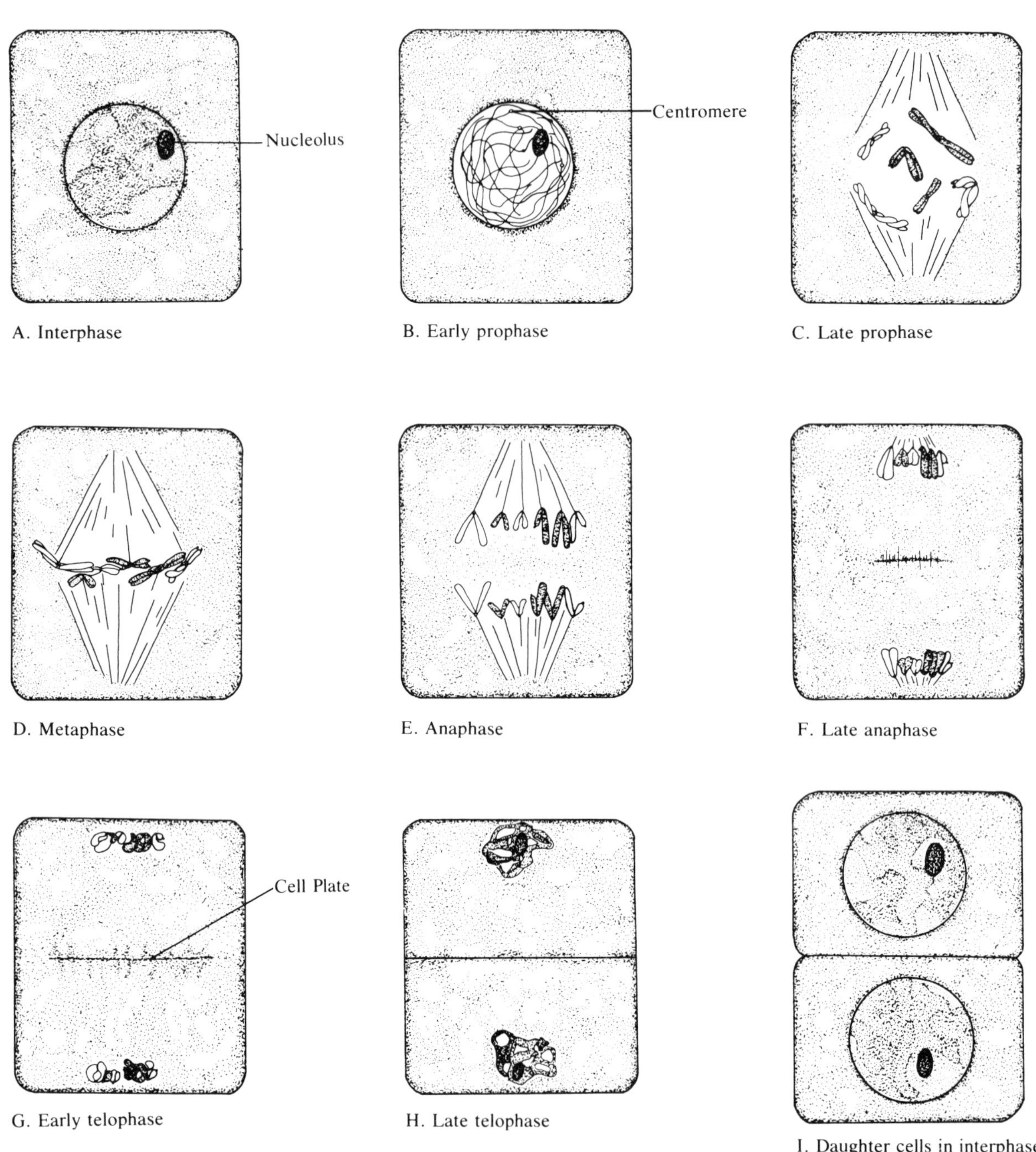

Figure 12.1. Plant mitosis.

DRAWINGS:

Prophase
Metaphase
Plant
Animal
Plant
Animal

Anaphase
Plant
Animal

Telophase
Interphase
Plant
Animal
Plant
Animal

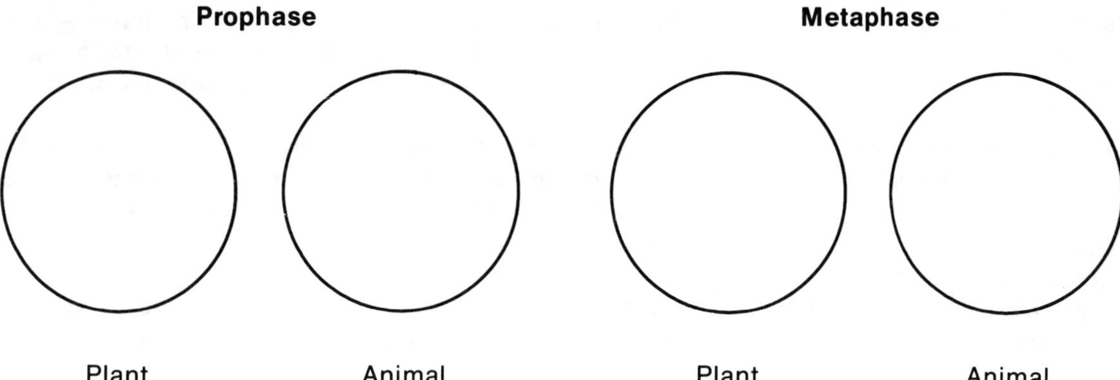

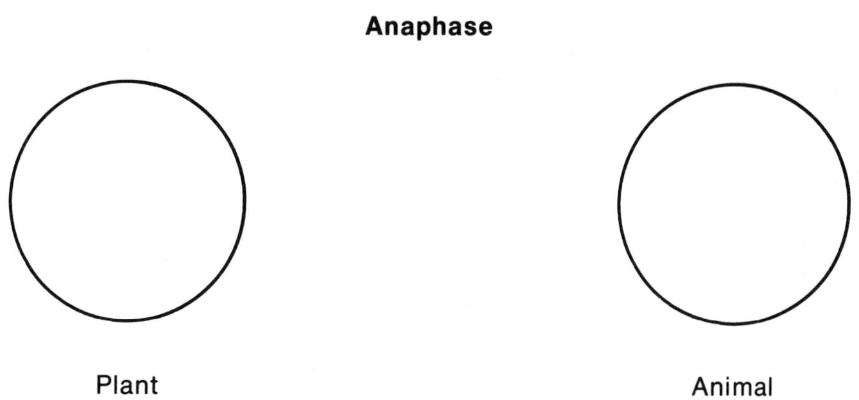

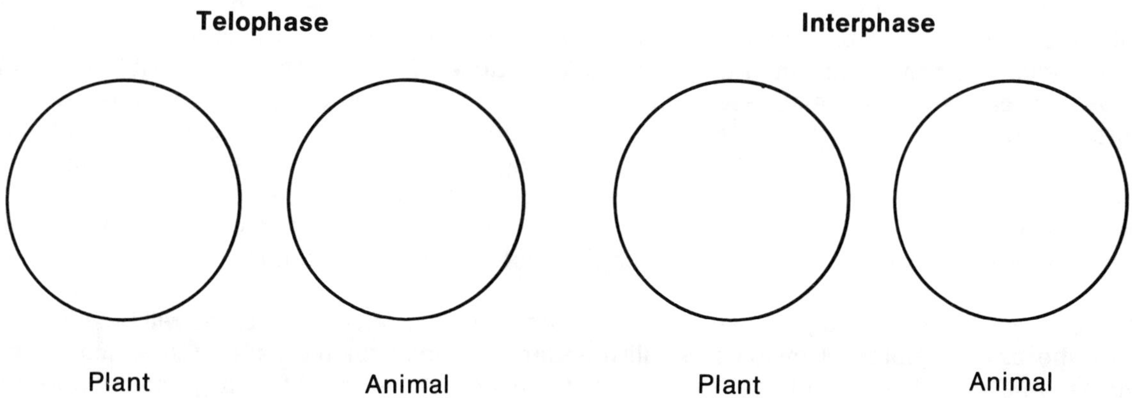

MEIOSIS

Cells from multicelled organisms may be divided into two general types: (1) **somatic** — those composing non-reproductive tissues, and (2) **germinal** — those concerned with gamete formation. In an organism, somatic and **early** germinal cells contain the same number and kinds of chromosomes.

Gametes, or sex cells, are the result of a special type of cell division called **meiosis.** The meiotic process reduces the number of chromosomes so that each daughter cell receives exactly one-half the original chromosome number (one member of each chomosome pair). Thus, the **diploid** *(2n)* parent cell, through meiosis, gives rise to **haploid** *(n)* daughter cells. Meiosis is usually a continuous sequence of events involving two successive divisions, during which chromosome replication occurs only once. Meiosis I is essentially a reduction division, while Meiosis II is a mitotic division. **The result of the complete process is four cells, each with the reduced (haploid) number of chromosomes.**

Gametogenesis is the series of changes, including meiotic division, which germ cells undergo prior to functioning as gametes in the process of reproduction. The general processes involved (such as chromosome behavior) are similar in both plants and animals, although differences are present in the immediate results. Gametogenesis in animals results in the formation of mature germ cells **(Figure 12.4).** In plants, asexual reproductive bodies **(spores)** are produced. Each spore is a part of the life cycle. It may develop into a pollen grain or embryo sac of a flowering plant, or into the leafy shoot of a moss. Ultimately, these spores will produce the germ cells. It is this sequence of events which results in an alternation of generations, commonly associated with plants, but rarely with animals. Meiosis and fertilization maintain a constant chromosome number in sexual reproduction in both plants and animals.

Sexual reproduction, showing the relationship of the processes of mitosis and meiosis in the animal's life cycle, is illustrated in **Figure 12.2.** Mitosis produces cells which contain the same number of chromosomes as the parent cell. Meiosis results in cells which contain half as many chromosomes as the parent cell. **Fertilization,** the union of a male and a female gamete to produce the zygote or fertilized egg, restores the original chromosome number of the parent cell in the offspring.

A knowledge of the differences and similarities of mitosis and meiosis is essential to understanding certain aspects of genetics and development which will be covered later in the course. Meiosis differs from mitosis primarily in chromosome behavior.

In the diploid cell, chromosomes are present in pairs **(homologous pairs).** In mitosis, each chromosome is independent. In meiosis, homologous chromosomes become aligned very closely, but not fused. This exact alignment is called **synapsis.** At this time, by the process known as **crossing-over,** exchange of sections of the chromosomes may occur between members of a pair.

Another possibility for producing different distributions of the chromosomal material in the daughter cells is **random assortment.** This occurs because of the number of possible ways the chromosomes can line up in Metaphase I. Thus the single member of each chromosome pair which a daughter cell will receive is strictly a matter of chance distribution. For example, in a cell with two pairs of chromosomes, both paternal chromosomes might go to the same cell and both maternal chromosomes to the other, or each cell could receive one maternal and one paternal chromosome. Random distribution of maternal and paternal chromosomes also occurs in organisms with many pairs of chromosomes, producing correspondingly greater numbers of different chromosome combinations in the resulting eggs or sperm.

In the following exercise, diagrams and models having one pair of chromosomes will be utilized to illustrate the basic features of meiosis. Small numbers are used for the sake of simplicity, but it should become readily apparent that more chromosomes per cell result in a greater number of possible combinations of genetic material in the resulting daughter cells.

Study the diagrammed sequence in **Figure 12.3.** Note the important features of each phase in the process. Also refer to **Figure 12.4** and your text for additional details.

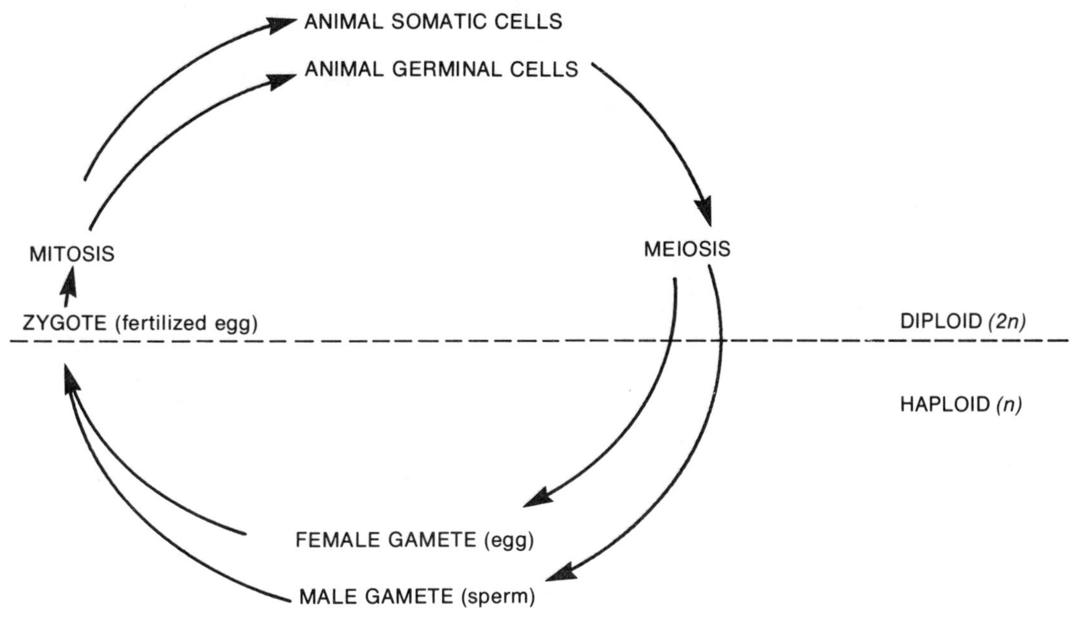

Figure 12.2. Life cycle of an animal with sexual reproduction.

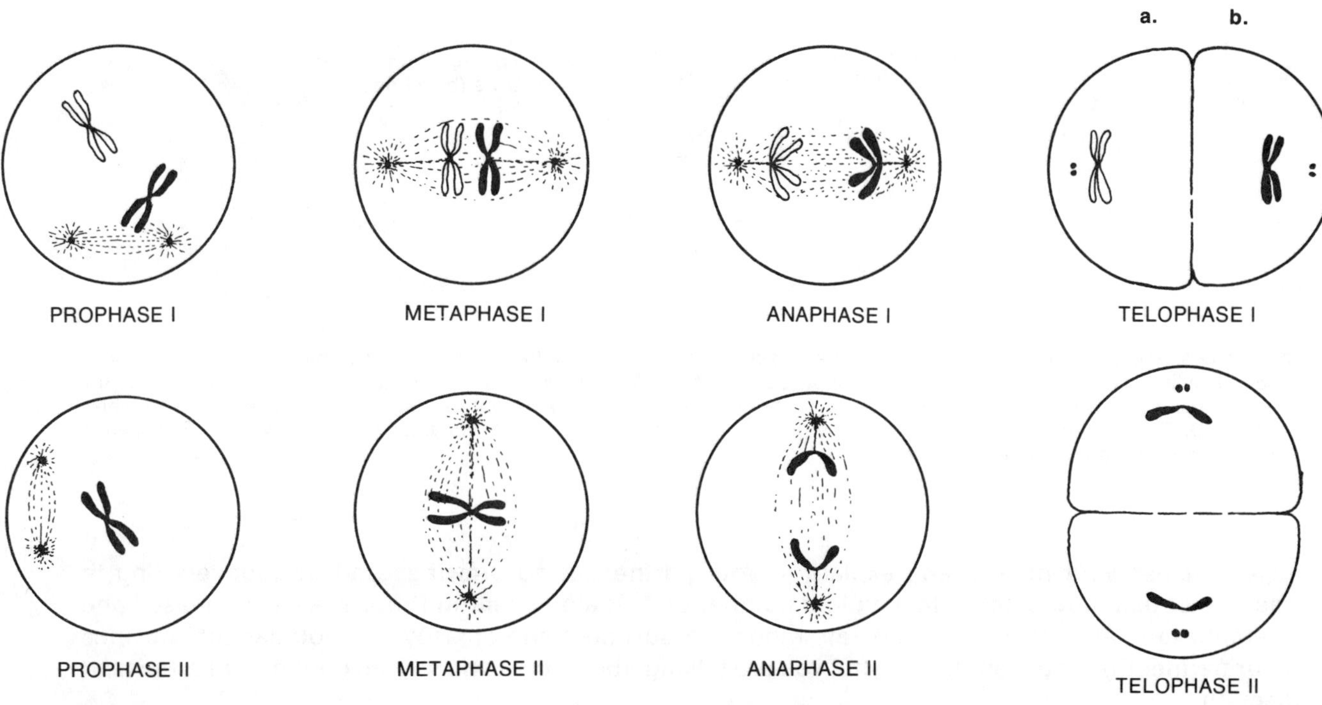

Figure 12.3. Meiosis of a cell containing a single pair of chromosomes. For simplicity, only cell (b) is followed through the second series of divisions, although both undergo the same steps.

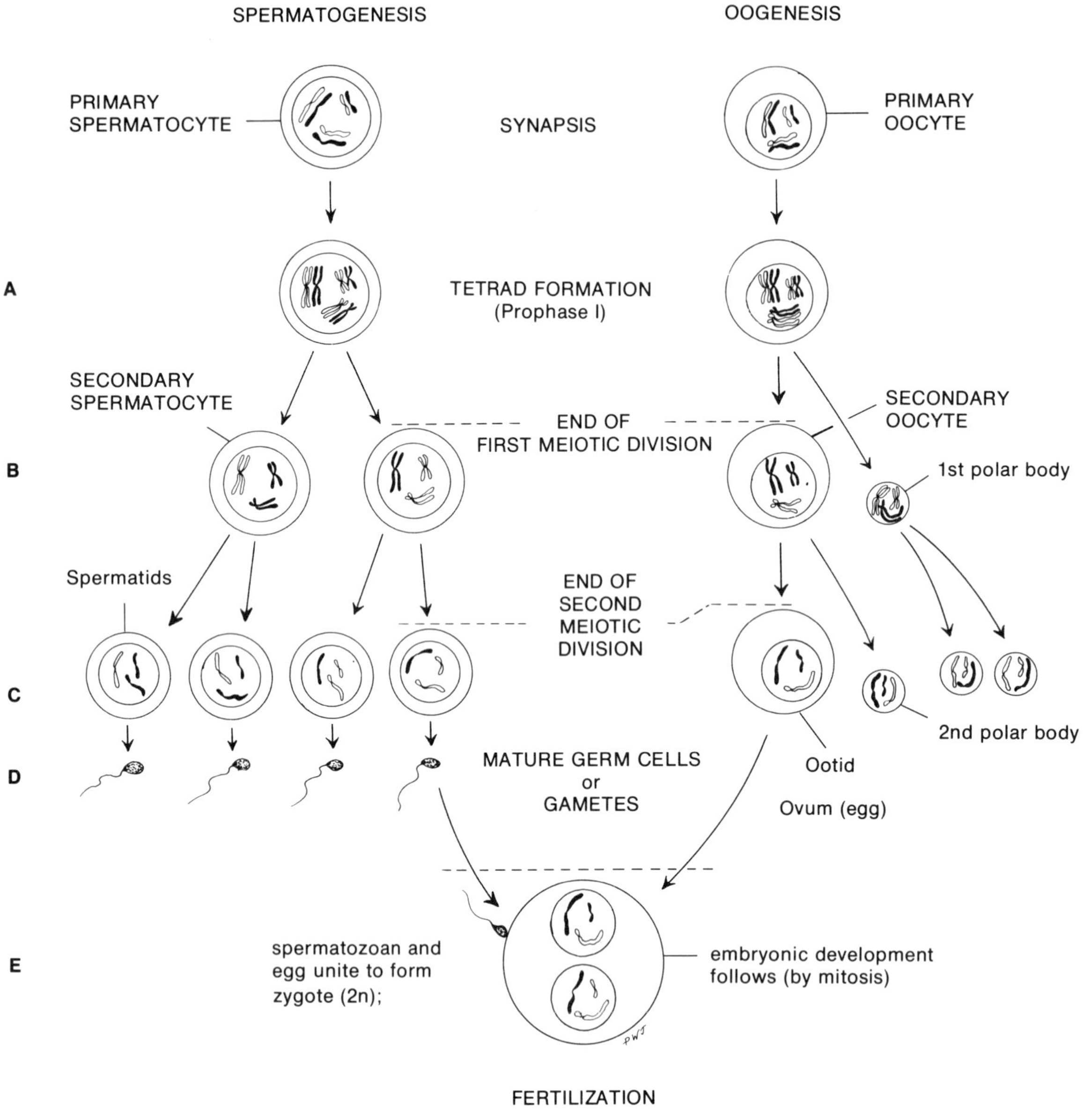

Figure 12.4. Animal meiosis. **A.** Chromosomes have assumed a doubled structure. **B.** At the end of the first meiotic division, 2 haploid cells result. Chromosomes are still composed of 2 chromatids attached to a single centromere. **C.** At the end of the second meiotic division, 4 haploid cells result. In the female, 3 of these are nonfunctional polar bodies. This completes meiosis. **D.** The 4 spermatids and the ootid undergo differentiation to form sperm and an ovum respectively. **E.** Fertilization restores the diploid number and mitosis begins.

Now, close your manual and explain to your partner, using drawings and without referring to manual or text, how a cell with a diploid number of 4 would appear in Prophase I, Metaphase I and Metaphase II. Also define and illustrate random assortment and crossover. If you cannot convince your partner that you really know what you are talking about, then you should continue with the next section.

At a quick glance the process of meiosis appears quite simple. **Most** students, however, experience difficulty in **understanding** the process (as evidenced by quiz grades). Therefore, we strongly urge you to use the tinkertoys — they really do help.

Chromosome Models

Work in pairs or even as a table unit of four if desired. Go through the following steps with the models, using the diagrams where needed for reference, and then you should be able to convince your partner.

1. Prepare models of two pairs of chromosomes. Take two long sections of pipe cleaners and place 10 white beads on each. Fasten the two together with a short section of pipe cleaner (to represent the centromere holding the two chromatids together). Place the centromere 3 beads from one end. Repeat with two other long sections, using blue beads. Then take two short sections of pipe cleaner and place 5 white beads on each. Again fasten the two together, placing the centromere 2 beads from one end. Repeat with two other short sections, using blue beads.
2. Mark off a large circle (approximately 8 inches in diameter) on a piece of paper. Arrange the 4 model chromosomes in the circle as they would appear in prophase I. Sketch their arrangement in the appropriate circle on the page for drawings. Label chromatid and centromere on one of them.
3. If crossing-over occurred, there would be an exchange of a section or sections of the chromosome between members of the homologous pair. Change the beads on the two large chromosomes to illustrate crossover in the segment represented by the last four beads of one chromatid. Assume that only one chromatid of each chromosome is involved. Use the long arm of the chromatid. Draw the resulting chromosomes. Remember that crossing-over is possible in a number of places, and the length and number of segments exchanged will vary.
4. In your large circle draw an equator and centrioles and arrange the four chromosomes on the equator as they would be in metaphase I. Use colors or some difference in designation to distinguish between maternal and paternal chromosomes. Draw both possible arrangements. This illustrates random assortment—the position chromosomes assume is strictly a matter of chance. How does this arrangement on the equator differ from that which would occur in metaphase of mitosis?
5. Centromeres do not divide at this time. One member of each homologous pair (consisting of two chromatids attached by a centromere) moves to each pole. Draw this cell as it would appear in anaphase I. For simplicity, draw only one possible arrangement based on one chosen from metaphase I.
6. Telophase I will follow and two cells will be formed — each with the haploid number.
7. Use each half of the circle previously drawn to represent a daughter cell. Arrange the chromosomes as they would be in prophase II and draw.
8. Draw an equator and centrioles in each daughter cell. Arrange the chromosomes as they would be in metaphase II and draw them.
9. In anaphase II, centromeres divide (remove the piece of wire representing the centromere), and each chromatid is now a chromosome since it has its own centromere. Chromosomes move to opposite poles. Draw these cells.
10. Telophase II occurs with each cell dividing, resulting in the formation of four cells. Draw the chromosomes which would occur in these cells.
11. If the chromosomes had randomly assorted in the other possible arrangement during metaphase I, what would the four daughter cells look like?
12. Study the meiosis models on display in the lab.
13. Film loops which show mitosis and meiosis are available for you to study.
14. The process of animal meiosis is summarized in **Figure 12.4.**

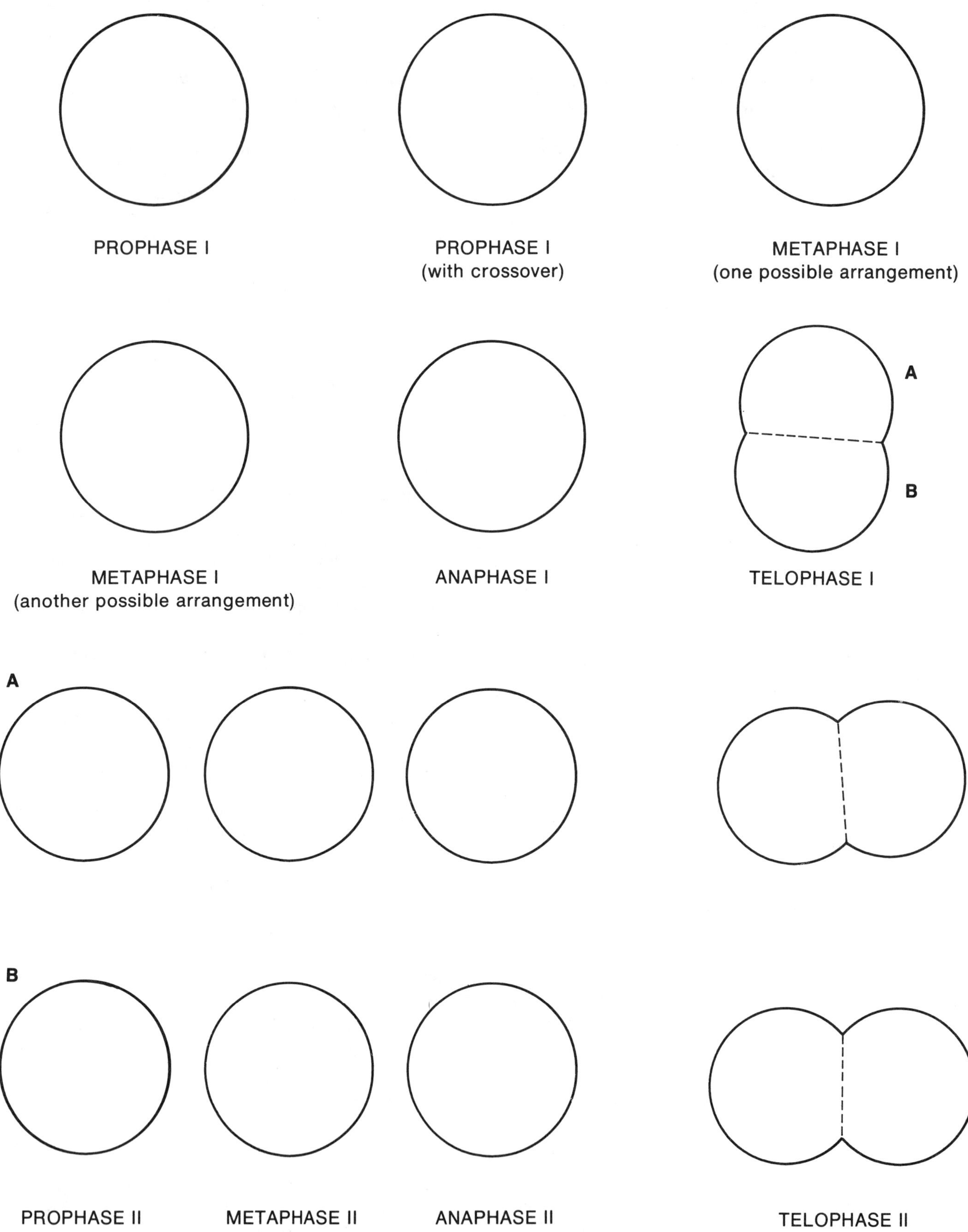
PROPHASE I
PROPHASE I
(with crossover)
METAPHASE I
(one possible arrangement)
METAPHASE I
(another possible arrangement)
ANAPHASE I
A
B
TELOPHASE I
A
B
PROPHASE II
METAPHASE II
ANAPHASE II
TELOPHASE II

QUESTIONS

1. Complete the chart to show likenesses and differences.

	Mitosis	**Meiosis**
Occurs in what types of cells	__________	__________
Number of cells produced	__________	__________
Type of cells produced	__________	__________
Stage when centromere divides	__________	__________
Number of planes on which chromosomes line up in metaphase and metaphase I	__________	__________
final chromosome number	__________	__________

2. If a liver cell of a certain animal has 38 chromosomes, how many would be found in the following cells of the same animal?

skin__________ lung__________ sperm__________ egg__________ liver cell from third generation__________ egg cell from second generation__________.

3. What two processes occur in meiosis which can produce variation between offspring?__________

____

4. What is the purpose of meiosis?

5. **Figure 12.1** illustrates plant mitosis. What would be necessary to convert this to an illustration of animal mitosis?

6. One diagram represents meiosis and one mitosis in cells with three pairs of chromosomes. Identify the one representing meiosis.
Does it represent metaphase I or II?__________Why?

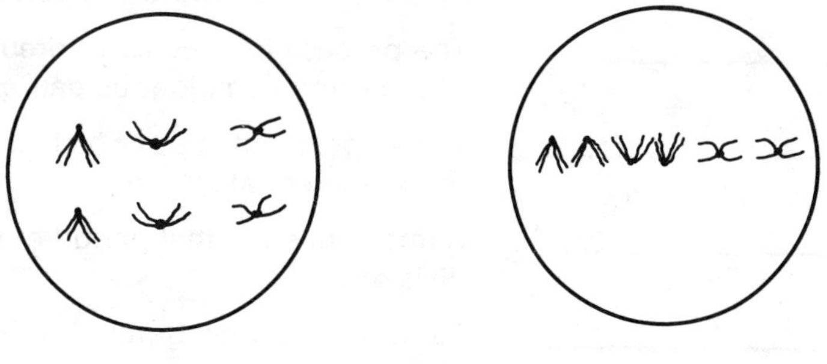

7. Is the diagram representative of prophase I of meiosis or prophase of mitosis?___________

Why? ____

In the blank circle draw the diagram the way it would look if it represented the other possibility.

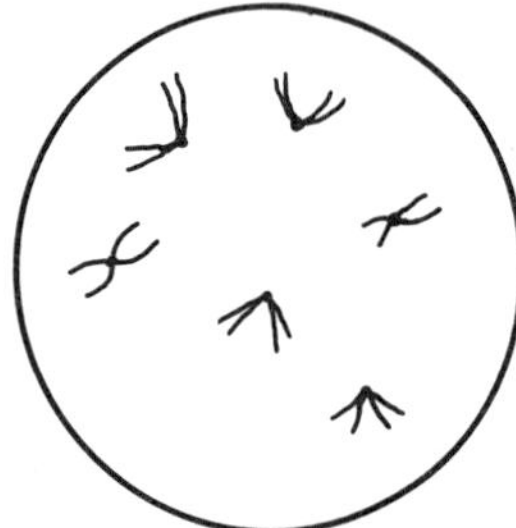

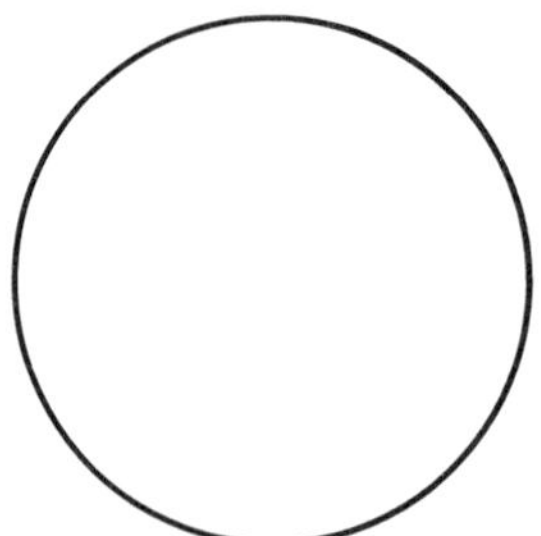

Building Your Biological Vocabulary

______________________	The *n* number of chromosomes which consists of one member of each homologous pair of chromosomes.
______________________	An animal cell organelle which moves to the poles of the spindle by late prophase.
______________________	The stage of mitosis at which chromosomes **begin** to move toward the poles.
______________________	The process in meiosis whereby chromosomes of the same homologous pair exchange genes.
______________________	The stage of mitosis at which chromosomes line up on the equator.
______________________	Name applied to that period of time prior to cell division.
______________________	Pairing of homologous chromosomes in prophase I.
______________________	A type of cell division which produces daughter cells containing the same chromosome number as the parent cell.
______________________	The point on the chromosome where spindle fibers attach.
______________________	A type of cell division which results in haploid gametes.

Glossary

aster — structure of radiating fibers which appears during prophase in animal cells, contains centrioles.
cell plate — the new cell wall which divides the cytoplasm of daughter plant cells.
centriole — cytoplasmic organelle which has tubular structure similar to cilia.

centromere — region on chromosome at which spindle fiber is attached during mitosis.
chromatid — one of the arms of a chromosome held together by a centromere during prophase. The chromatid contains one copy of the genetic material which was duplicated during interphase.
chromatin — the hereditary material of the cell which appears as scattered granular material during interphase.
chromosome — one of the rod-like bodies which contains the genetic material of the cell during mitosis. All of the material attached to a single centromere is considered to be a single chromosome.
cleavage furrow — the region of the cell in which the cytoplasm of the new daughter cells is pinched apart during cytokinesis in animal cells.
cytokinesis — cytoplasmic division.
diploid — chromosome state in which each type of chromosome (with the possible exception of the sex chromosome) is represented twice *(2n)*.
haploid — state in which each type of chromosome is represented once; the cell contains a single set of unpaired chromosomes. This is the condition found in gametes and is represented by *n*.
homologous chromosomes — similar chromosomes forming a pair; each member of this pair contains hereditary information (genes) affecting the same characteristics or traits.
spindle — biconical structure of fibers running between opposite poles of the cell. Each centromere becomes attached to a spindle fiber.

LABORATORY 13
PATTERNS OF INHERITANCE

OBJECTIVE: To apply the principles of meiosis to predict types and ratios of offspring resulting from specific crosses.

Living organisms produce offspring like themselves. The cytoplasm and nucleus of sperm and egg combine in the process of **fertilization** to form the zygote. From this single cell will arise a many-celled organism with characteristics derived from both parents.

During cell division chromatin material in the cell nucleus condenses into a number of rod-like chromosomes. By examination of chromosome behavior, researchers learned that the determinants of heredity were arranged in linear fashion on the chromosome like beads on a string. Prior to knowledge of the chemical model involved, the term **gene** was coined to designate these unknown hereditary determiners. The term gene has a certain vagueness but will be used here as representing that portion of the chromosome responsible for determining some aspect of the organism containing it. We will redefine the term at the completion of the next exercise.

In normal body or **somatic** cells, chromosomes are present in **homologous pairs.** Chromosomes of an homologous pair look alike, stain alike and contain genes governing the same traits. Each gene site or **locus** pertains to the same characteristic. In **Figure 13.1,** locus (a) governs plant height. Genes at this locus on each chromosome of the homologous pair may be identical or different (as shown here), but all genes found at this specific site must contain information pertaining to height. Either member of such a pair of genes is known as an **allele.** If both genes of a pair are alike, the organism possessing them is said to be **homozygous** for that gene. If the members of the pair differ from each other, the organism is said to be **heterozygous** for that trait. In **Figure 13.1,** this plant would be homozygous for flower color and leaf shape and heterozygous for height and leaf surface.

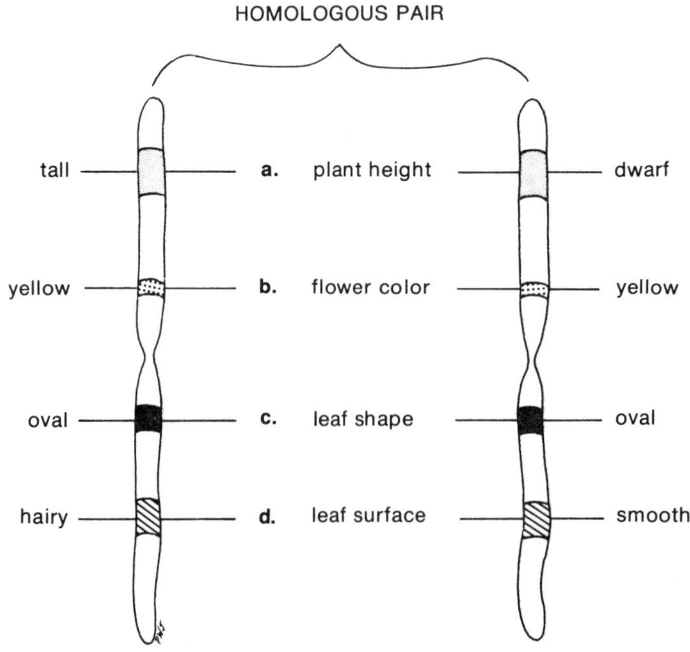

Figure 13.1. A pair of homologous chromosomes showing four different loci.

Let us use these terms and add additional terms as we look at the processes of meiosis and fertilization. In most organisms cells contain many chromosomes and each chromosome contains many genes. In studying genetic problems you look **only** at the homologous pair(s) of chromosomes carrying the particular genes being studied. With this in mind, the cell from a male, tall plant homozygous for the tall gene would appear as shown on the left. On the right is a cell from a female dwarf plant homozygous for the dwarf gene. Tall and dwarf are **phenotypes.** TT and tt are **genotypes.**

1. In the circle draw the chromosome picture which would be seen in sperm cells in pollen produced by the male plant described previously.

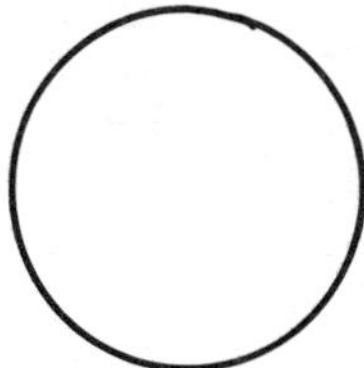

2. If you have two chromosomes in this cell — you goofed! Meiosis would result in only one member of each pair going into the gametes formed. All sperm produced by this male would contain only one gene for tall.
3. In the circle draw the chromosome picture which would result if the female plant described previously produced gametes. (Again, meiosis would lead to only a single gene for dwarf in each gamete produced.)

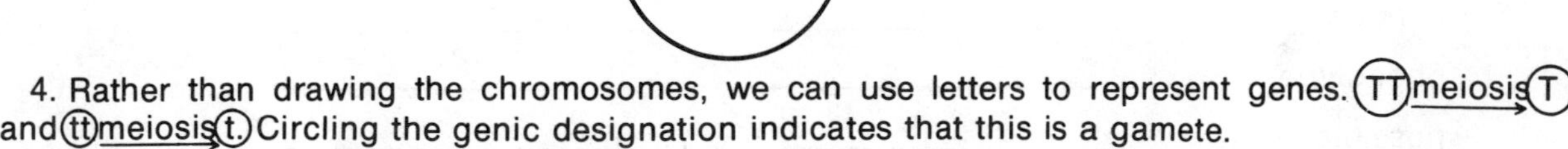

4. Rather than drawing the chromosomes, we can use letters to represent genes. (TT) meiosis → (T) and (tt) meiosis → (t) Circling the genic designation indicates that this is a gamete.
5. Now record the genotype of offspring resulting from fertilization of this egg by this sperm.

____

6. You should have restored the diploid number in the offspring, and all offspring produced by this mating would have the genotype Tt.

A capital letter is used as an abbreviation for an allele which will be **expressed** in the organism (that is, a **dominant** gene). The lower case **of the same letter** represents the **recessive** allele.

Monohybrid Cross

In organisms that produce large numbers of offspring, one may use the known happenings of meiosis to predict possible types of offspring and ratios in which they may occur. Problems arise when studying organisms which produce only a few offspring from a single mating. To determine patterns of inheritance would require that these parents be mated many times or that many similar parents be mated. **No one mating that produces few young can be expected to show all the possible genetic combinations in the offspring.** Consequently, we determine the **probability** that a given combination of alleles will appear in an individual.

1. Now — let us take our terminology and apply it to predict certain information about the offspring of a particular cross. The class will work together in answering the following problems.

2. **Problem:** A male grey mouse (homozygous for the dominant allele G) and a female black mouse (homozygous for the recessive allele g) are mated. What genotype and phenotype would you predict for the offspring of this cross? Use the given information to complete the following chart.

	Parent ♂	♀
phenotype	______________	______________
genotype	______________	______________
possible types of gametes	______________	______________

fertilization
↓
zygote

genotype ______________

phenotype ______________

3. Let us go a step further. The original mating represented the **P** or **parental generation.** The offspring represent the F_1 or **first filial generation.**

4. What different genotypes and phenotypes could result if two of the offspring from the cross above mated with each other? $F_1 \times F_1 \longrightarrow F_2$ **(second filial generation).** A good way to be certain you include all possibilities is to utilize a **Punnett Square.** Complete the following.

		♂	♀
F_1	genotype	______________	______________
	phenotype	______________	______________
	possible gamete types	______________	______________

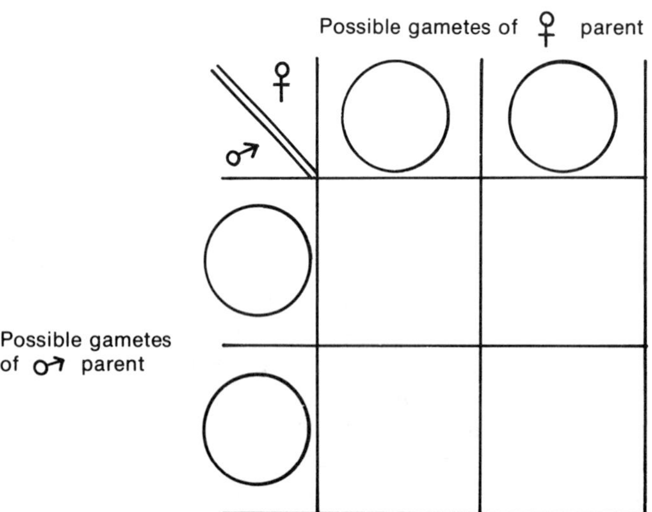

The four boxes contain the **possible** genotypes of individual offspring resulting from combination of that sperm and egg. List these genotypes and the phenotype of each.

F_2	**genotype**	**phenotype**
	____________	____________
	____________	____________
	____________	____________

Predicted genotypic ratio ____________

Predicted phenotypic ratio ____________

Sample Problem

1. The lab instructor will furnish each table with a genetics problem. Each student should determine the phenotypes, genotypes and gamete types produced for each parent in this problem. Calculate offspring genotypes and phenotypes and the possible ratios and enter this information on page 118.

2. Now, let's run a trial experiment and compare these predicted results to some experimental data.

3. **Use the same problem.** Let the blue beads represent the dominant trait and the white beads the recessive trait. Each bead will represent a gamete carrying the specified trait.

4. Place an equal number of beads for each gamete type (10 will be enough) into the beaker marked ♀. (For example, **if** an individual could produce two types of gametes you would need 10 blue and 10 white beads in the same container.) Also, count out the proper number and color of beads and place in the ♂ container.

5. Mix the beads in the ♂ container and let one student reach in and draw out a single bead, **sight unseen.** Remember, the process is random. This represents a sperm. Record the gene designation in column 1, page 118. **Return the bead to the container to maintain the gene pool constant.** Draw out a bead from the other container. This represents an egg which will be fertilized by the sperm. Record the gene designation of this egg in column 2, page 118, and then record the genotype and phenotype of the individual which would develop from this zygote.

6. Repeat this procedure twelve times. Then calculate the ratio obtained and compare it to the predicted ratio for this type of cross.

7. When everyone has completed the data, the instructor and the class will go over some of these problems in a class discussion.

8. Conclusions based on ratio predicted and ratio obtained.

Dihybrid Cross

The reasoning used in the previous problem may also be applied to **dihybrid crosses.** These are crosses between individuals that differ with respect to two specified gene pairs. In the problems which you will consider, genes for one trait are found on one chromosome pair and genes of the other trait are on a different chromosome pair.

Sperm	Egg	Zygote	
		Genotype	Phenotype

	Genotypic	**Phenotypic**
Ratio Obtained:	______________	______________
Predicted Ratio: (calculated from problem)	______________	______________

Conclusions:

Dihybrid Cross (cont.)

1. **Problem:** A male grey, long-haired mouse (homozygous for both G and L) and a female black, short-haired mouse (homozygous for g and l) are mated. What genotypes and phenotypes would you expect to find in their offspring?

	♂	♀
phenotype	grey, long-haired	______________
genotype	GGLL	______________
possible types of gametes	(GL)	______________

fertilization
↓
zygote

genotype ______________
phenotype ______________

2. Now, what genotypes and phenotypes would be expected in the F_2 generation? Use the Punnett square, but this time you will need 16 squares instead of just four. Why? There are four possibilities for types of gametes which could be produced by one of these F_1 individuals. These four are:

List these four types across the top of the Punnett square. Then list the four types produced by the other parent vertically at the left side of the Punnett square. Combine these to obtain possible genotypes of the F_2 generation. (Genotypes are located inside the 16 boxes). From this information,

what type of genotypic ratio would you predict? ____________________

phenotypic ratio? ____________________

Human Chromosomes

Observe the demonstration slide of a human chromosome preparation. Note chromatids, centromere, differences in size and shape, etc. It is sometimes possible to correlate chromosome appearance or numbers with particular phenotypic traits (as in some forms of Down's syndrome). See if you can identify any of these chromosomes (by number) by comparing them with the chart.

Sex-Linkage

In man and some other organisms, a pair of chromosomes **(sex chromosomes)** normally determines the sex of the individual. The remaining pairs of chromosomes are known as **autosomes.** Man has two kinds of sex chromosomes, the X and the Y. When an individual receives an XX chromosomal constitution, a female results, and the XY combination produces a male. Normally the YY combination does not occur. Types of gametes produced by a man are X and Y, and by a female X only.

Characteristics associated with the X chromosomes are said to be **sex-linked,** and ratios obtained in crosses involving sex-linked characteristics will differ from those obtained in autosomal characteristics. Also, note that **since the male carries only one X, whatever allele is present on this X is present only once and will be expressed even if recessive.** The Y carries few characteristics and for our purposes can be assumed to carry none.

Hemophilia is a disease involving excessive bleeding and is associated with a recessive sex-linked characteristic (X^H being the normal allele and X^h the recessive allele). Supply the following.

phenotype	**genotype**	**possible gametes**
normal female	________	____________
normal male	________	____________
hemophiliac male	________	____________

Multiple Alleles

1. I^A, I^B and i are allelic genes for human blood type leading to the blood groups A, B, AB and O. An individual receives only two of these alleles (one from each parent). I^A and I^B are dominant over i, but not over each other.

Blood Group	Possible Genotypes
Yours (if known) ____	____________
A	____________
B	____________
AB	____________
O	____________

2. Should you desire, materials are available to determine your blood group. Remember that an antigen (on the cell) and an antibody (in the typing sera) will clump or agglutinate if they are the same (for example, A with anti-A). Blood group A means that A antigens are present, and B that B antigens are present. AB has both antigens and O has none. With this in mind, complete the chart as to whether or not agglutination would occur with each combination of antigen and antibody.

Cell Antigens	Anti-A Serum	Anti-B Serum	Cell Antigens	Anti-A Serum	Anti-B Serum
A			AB		
B			O		

3. **How to Determine Blood Group.** With a wax pencil mark a slide as shown in **Figure 13.3.** Thoroughly clean the tip of the ring finger with an alcohol soaked cotton ball. ''X'' on the diagram in **Figure 13.3** marks the best spot to stick. Remove a **new** lancet from its package and use it to prick your finger (or get your partner to do it if you are cowardly). You need to get two drops of blood so stick it hard enough the first time. **Discard the lancet and do not under any circumstances use this lancet on another person.** Why? Place a drop of blood in each circle on the slide. Add a drop of anti-A to circle A and anti-B to circle B. Use a piece of an applicator stick to mix anti-A and the blood. Then **use the other end** of the stick to mix the blood and the anti-B. Tilt the slide back and forth gently for a minute or two and observe for agglutination or clumping of the cells. If no clumping occurs, the mixture will remain homogeneous. You might want to look microscopically at a mixture that does show agglutination. Should this sort of clumping occur in an animal's blood vessels, can you foresee any possible problems in circulation? ____________

in excretion? ____________

Record your results on the diagram in **Figure 13.3.**

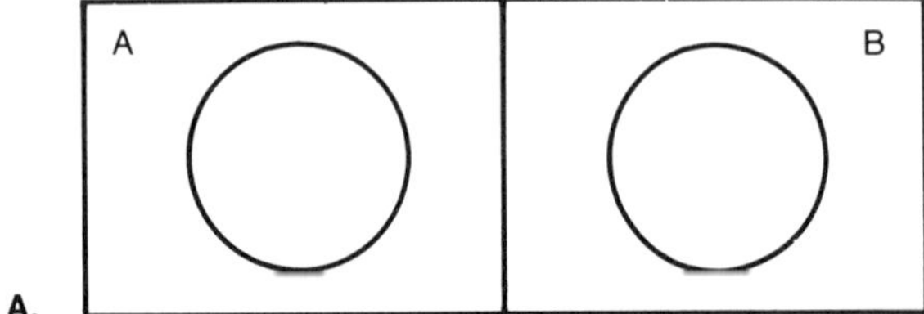

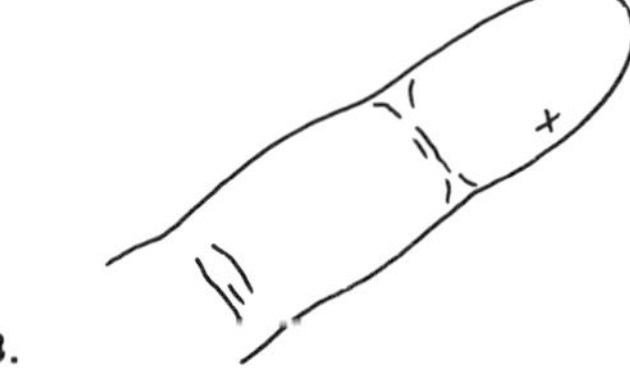

Figure 13.3. A, Slide preparation for determining blood group; **B,** Diagram indicating best area to prick with the lancet.

4. A man with group A blood marries a woman with group B.

Possible gametes: of the man ________________

of the woman ________________

Possible blood groups of their children:

5. A man with group B blood (whose mother had group O) marries a woman with group O.

Possible gametes: of the man ________________

of the woman ________________

Possible blood groups of their children:

Selected Problems

The following are typical types of problems which you should be able to work. Answers are given at the end of the exercise, but **do not** look at the answers until you have tried to work the problems. Use the following information in working the problems.

Dominant Character	Recessive Character
normal pigmentation	albino
black (guinea pig)	white
polled (cattle)	horns
short hair (guinea pig)	long hair
rough coat (guinea pig)	smooth
normal vision	color-blind
normal clotting	hemophilia
purple flowers	white
tall plants	short

1. Mr. Jackson, an albino, marries Miss Williams (who has no record of albinism in her family). With respect to albinism, what kind of children may they expect to have? Children's genotype?
2. Twenty years later one of the Jackson children marries an albino individual. Give expected genotypic and phenotypic ratios among their children.
3. Tarzan and Jane marry. Both had one normally pigmented parent and one albino parent. What phenotypic and genotypic ratios would you predict for their children?
4. Tarzan and Jane have 8 children, all normally pigmented. Does this agree with your predictions in the previous problem? If not, explain the situation.
5. A rancher breeds a polled cow to a polled bull and the resulting calf has horns. Give the genotypes of the three animals.
6. A homozygous black female guinea pig with long hair is bred to a male guinea pig with white, long hair. What phenotypic ratio would you predict as a possibility among the F_1 and F_2 offspring?
7. You want to breed a white male guinea pig, homozygous for rough coat to a female black guinea pig with a smooth coat. The female's mother had white hair and a rough coat. Give the genotypes of these three individuals and determine types of offspring the cross could produce.
8. Homozygous tall plants producing white flowers are crossed with homozygous short plants with purple flowers. The F_1 generation is then selfed or crossed with each other. Give phenotypic ratios expected in the F_2 generation.
9. Alice has normal vision, but her father was color-blind. She marries George, who is color-blind. What will be the phenotypic ratio among their children? The color-blind gene is located on the X or sex chromosome.
10. Betty is a carrier for hemophilia. She marries a man who has normal blood clotting time. What proportion of their sons will have hemophilia? Could they have a hemophiliac daughter? Remember that hemophilia is a sex-linked characteristic.

Answers to Genetics Problems

1. normally pigmented heterozygotes; Aa
2. 1 Aa: 1 aa; half normal: half albino
3. 3 normal: 1 albino; 1 AA: 2Aa: 1 aa
4. No; the process is random and ratios hold true for large numbers of offspring
5. cow: Hh; bull: Hh; calf: hh
6. F_1: all black, long hair; F_2: 3 black, long hair: 1 white, long hair
7. Male: wwSS; female: Wwss; mother: wwSs; offspring half black with rough coats: half white with rough coats
8. 9 tall purple: 3 tall white: 3 short purple: 1 short white
9. 1 normal female: 1 color-blind female: 1 normal male: 1 color-blind male
10. 50%; no

PATTERNS OF INHERITANCE CROSSWORD PUZZLE

Across

2. a haploid cell from a female
5. A___cross involves two different pairs of genes.
8. A person with group B blood has the B anti___on his red blood cells.
11. Red hair would be an example of a pheno___.
14. If both alleles for a given trait are identical, the organism is___at that locus.
18. A cow with the genotype Rr would be___zygous.
19. Genes are found at specific___on chromosomes.
20. Chromosomes other than the X and Y are referred to as___somes.
25. The gene for blood group A is an___of the gene for B blood group.
26. AA X aa → the first___generation
27. Karen (XX) marries Ted (XY). The daughter, Debbie, would have the genotype___.
29. Substance in the antisera which reacted with your blood was an anti___.
30. Chromosomes which are not sex chromosomes are auto___.
33. the unit of heredity
35. Eggs fertilized by sperm form___.
37. A fly homozygous for long wings (L) and ebony body (E) would produce gametes containing___.
39. If a single recessive gene is sufficient for a trait to be expressed in a male, the trait is probably sex___.
40. gametes of a rooster
41. Blood groups are an example of___alleles.
42. A cat with the genotype BB is homo___.
43. The F_1 generation is the offspring of the___generation.
44. A monohybrid___involves a single pair of genes.

Down

1. George Washington's genotype (sex chromosomes)
3. the genotype of a plant homozygous for the recessive gene "g"
4. LlWw describes the___type.
5. Rr and RR chickens are red, rr chickens are white. Gene R is___.
6. aabbcc plants are___zygous.
7. A typical dihybrid cross gives a 9:3:3:1 phenotypic___.
9. Long ears would be a___type.
10. Possible offspring from crosses may be predicted using a___.(two words)
12. A recessive sex-linked blood disease is known as hemo___.
13. the gamete of an individual with genotype EEee
15. Skunks with MmNN could produce gametes with MN (and, or) mN.
16. SS X ss would be a___hybrid cross.
17. the study of heredity
21. Red, tall, striped, etc. are descriptive of a particular___.
22. ___ual reproduction does not involve eggs and sperm.
23. A and B antibodies are found in human___.
24. Offspring of the parental generation are the___filial generation.
27. Gregor Mendel's genotype (sex chromosomes)
28. DDEE and DdEe are tall plants with round fruit. ddee are dwarf with elliptical fruit. d and e are___
31. Children of a cross are referred to as off___.
32. the genotype of a male hemophiliac
33. a gamete which could be produced by an elephant with the genotype GgEe
34. A, B and O are human blood___.
36. Haploid cells involved in fertilization are called___.
38. A 2n cell containing genes for E and R, would have the genotype___.

(See page 208 for the correct answers.)

PATTERNS OF INHERITANCE CROSSWORD PUZZLE

LABORATORY 14

DNA STRUCTURE AND PROTEIN SYNTHESIS

Note: It is suggested that paper models (pages 133-137) be cut apart prior to the lab period, as all working time will be needed to complete the exercise. They may be stored in the envelope at the back of the manual.

OBJECTIVES: 1. To become familiar with the components and organization of the DNA molecule.
2. To observe how DNA structure is altered in various types of mutations.
3. To see how DNA structure codes for amino acid arrangement in proteins.
4. To observe some examples of how altered DNA is expressed.

Deoxyribonucleic acid (DNA), a large molecule found principally within the nucleus, contains information which determines the characteristics of the cell in which it is located.

Structurally, DNA is composed of two types of **nitrogen bases** — double-ringed **purines** (adenine and guanine) and single-ringed **pyrimidines** (cytosine and thymine), a five-carbon sugar (deoxyribose) and a phosphate group. The combination of nitrogen base, sugar and phosphate produces a unit known as a **nucleotide.** Long chains of these units are polynucleotides. A DNA molecule may contain as few as a hundred nucleotide pairs or as many as several thousand. A segment of this strand functions as a gene, the unit of heredity.

In DNA it has been shown that for each nucleotide containing adenine (A), there is one with thymine (T). The same relationship occurs between guanine (G) and cytosine (C). These studies form a part of the basis for the formulation of the structure of DNA as two large polynucleotide chains hooked together by hydrogen bonds (H bonds) between the bases. This arrangement will produce a ladder-shaped structure. (See display models in the lab.) This ladder is then twisted to produce an alpha-helix with a spiral staircase appearance. The previous statement that the ratios of A:T and G:C are 1:1 is also used to support the theory that A bonds only to T and G bonds only to C. While the A:T and G:C ratios are equivalent in all double-stranded DNA, the ratio of AT:GC varies in different organisms.

Another important nucleic acid is **ribonucleic acid (RNA).** RNA contains uracil (U) instead of thymine, ribose rather than deoxyribose and exists in several forms which have different functions in the cell.

1. Study the structural formulas in **Figure 14.1.** In what way do ribose and deoxyribose differ? Draw a box around the area which differs in the two molecules.
2. In what major way do purines and pyrimidines differ from each other? ______________________

____

3. Note similarities between G and A.
4. Note similarities between C, T and U.

A Look at Spatial Configuration

1. Three dimensional models of the molecules shown in **Figure 14.1** are on display.
2. Following is a brief review of the models. Each colored ball represents a different atom.

carbon — black | hydrogen — white | oxygen — red
nitrogen — orange | phosphorus — purple

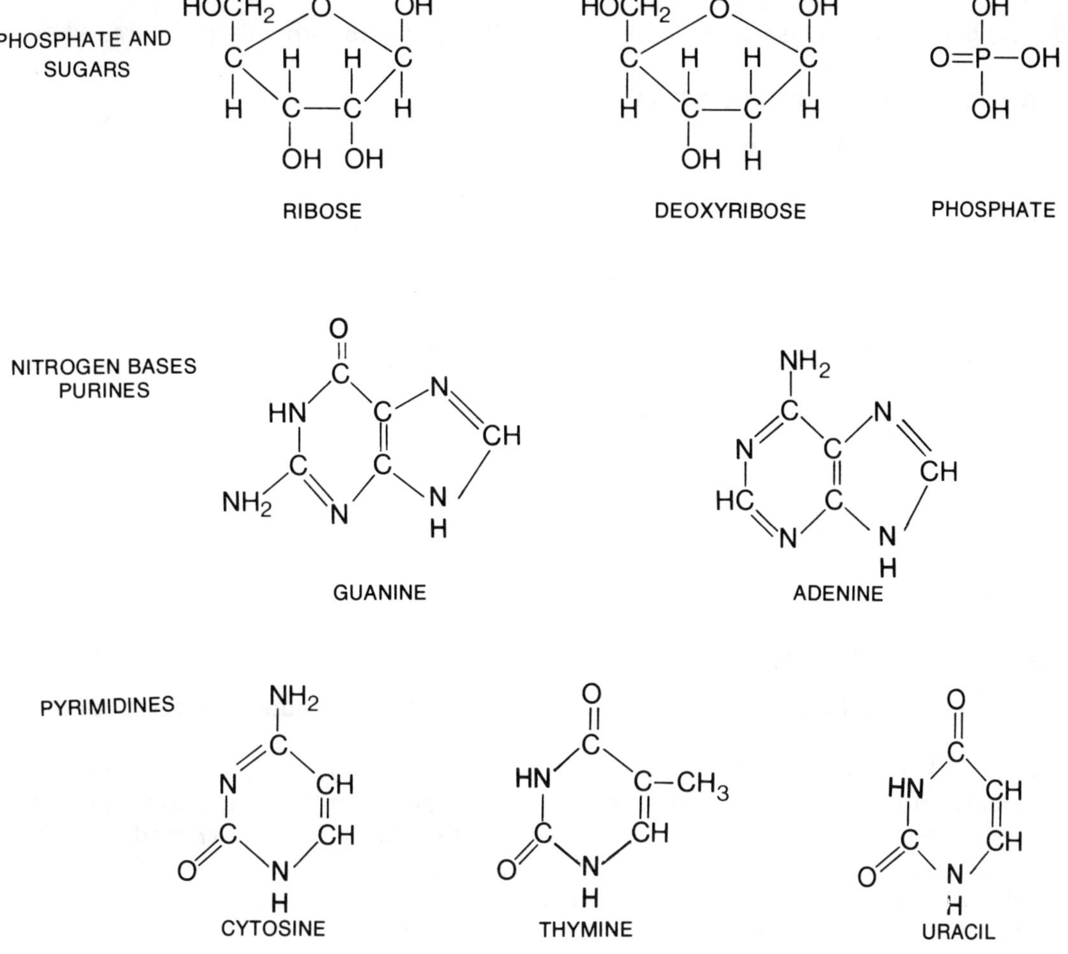

Figure 14.1. The structure of the phosphate group, sugars and nitrogen bases found in DNA and RNA.

3. Each spring represents a single shared pair of electrons, or chemical bond. The longer springs represent bonds between C — C, C = O (double bond) and C — N. Short springs are used for bonds between C — H, C — O (single bond), O — H and N — H. These differences correspond approximately with known atomic spacing in organic molecules. The number of holes in each model atom corresponds to the number of bonds it usually forms. Double bonds are formed by connecting the member atoms with two long springs.

4. Compare the models with the two-dimensional structural formulas **(Figure 14.1).** As you study the models note the difference in the **size** of purines and pyrimidines. Which one has a single ring structure?____________________ Which one a double ring structure?____________________

Draw an example of a purine and a pyrimidine (omitting atoms which project out from the main structure).

5. Observe the nucleotide model. A nucleotide consists of a nitrogen base bonded to a____________________ and a____________________ **(Figure 14.2).**

6. Study the double helix model. Observe how nucleotides fit together to form this structure. What molecules form the side pieces of this twisted ladder? ____________________

What molecules form the rungs of the ladder? ____________________

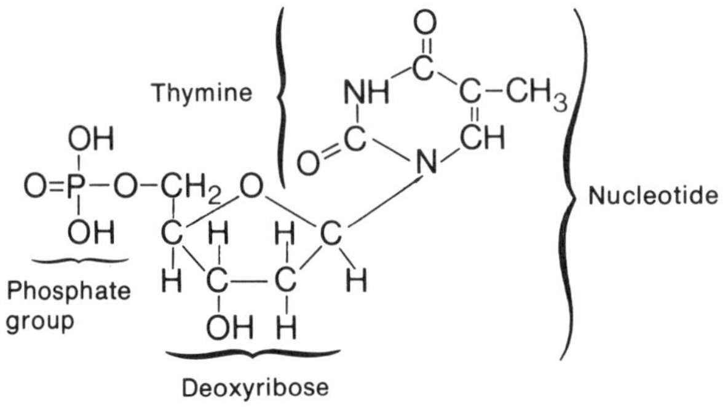

Thymine

Nucleotide

Phosphate group

Deoxyribose

Figure 14.2. Thymidylic acid, a thymine deoxyribonucleotide found in DNA.

DNA Structure — Paper Models

1. Models needed are found on pages 133 through 137. An envelope is included at the back of the manual for storage.

2. To facilitate your work, H atoms and -OH groups have been omitted at points where bonding will occur. Hydrogen atoms and some attached groups are also omitted from the ring forms, except where essential to show bonding. Carbons have also been omitted from rings. Paper models have limitations as they do not permit accurate spatial representation of bonds.

3. Observe the nitrogen bases (A, T, G, and C). Place the four different bases side by side and note the differences between them, as well as the similarities. Refer also to the structural formulas given earlier in the exercise.

4. Make four different nucleotides using the paper models. What molecules are necessary to make a nucleotide? Lines indicating bonds **must match** or the model is not assembled properly. Now attach the four nucleotides to each other by means of bonds from the phosphate group of one nucleotide to the sugar of the next. You now have a single strand of DNA containing four nucleotides. Note that you can arrange the bases in any sequence on your strand (reading from the top down) and all components will still fit.

5. If you joined the single strand you have just made to its matching or complementary strand to form a double stranded DNA, what nucleotides will be needed?

Your sequence of bases ___ ___ ___ ___

Complementary sequence ___ ___ ___ ___

6. The bases would bond to each other by means of hydrogen (H) bonds (usually represented by ||||||||). Individual H bonds are relatively weak, but a number of them together result in a stable molecule.

RNA Structure — Paper Models

1. An RNA molecule is a sequence of nucleotides formed as a complementary fit to one DNA strand. Only one DNA strand serves as a **template** or pattern for producing RNA. Remembering that RNA does not contain thymine, fill in the chart.

DNA base sequence (your paper model) ___ ___ ___ ___

Complementary RNA base sequence ___ ___ ___ ___

Complementary DNA base sequence ___ ___ ___ ___

RNA sequence formed on this strand ___ ___ ___ ___

2. You can see that the two RNA's are not the same. The base sequence produces the code or message contained in DNA and any variation in this sequence will produce a different code. Therefore, only one strand functions as a template for transfer of coded information.

Coding of Information

Three bases together comprise a **codon** or **triplet,** and each codon codes for an amino acid (AA). Using four different nitrogen bases, there are 4^3 or 64 possible triplets for coding. This is more than necessary for the 20 different amino acids. Thus, more than one codon may represent the same amino acid. This is known as a **degenerate code.** For example, both UUC and UUU code for phenylalanine, and both UCU and UCC code for serine.

What you have just read means that a sequence of three nitrogen bases in DNA will in some way cause the placement of a specific amino acid at a specific place in a protein chain. To see how this is accomplished let's look at the process of protein synthesis.

PROTEIN SYNTHESIS

Transcription

DNA remains in the nucleus. Proteins are formed in the cytoplasm. So — we need something to carry information from nucleus to cytoplasm. The messenger which will carry this information to the cytoplasm is formed by the process of **transcription.** Individual nucleotides free in the nucleus are brought to the DNA strand and lined up in such a way that complementary bases are paired.

By this process the sequence of bases coding for a specific protein transfers its information by forming a complementary strand of RNA. This RNA is called **messenger RNA (mRNA)** because it carries the message from the nuclear DNA to the site of protein synthesis in the cytoplasm.

Translation

1. Study the diagram of protein synthesis **(Figure 14.3)** as you read through the following steps.

2. The mRNA can now carry its message to the cytoplasm. Where does it go and what happens next? **Translation** will now occur — the building of a specific protein according to the specifications of the message. Protein manufacture occurs in ribosomes located in the ____________________.

3. Two additional types of RNA are required: (1) **ribosomal RNA (rRNA),** whose exact function is not clear, and (2) **transfer RNA (tRNA).** The tRNA attaches by means of a **high energy bond** to a specific amino acid and transfers this AA to the ribosome. Within the ribosome, mRNA and tRNA's come together, tRNA's release their amino acids, and amino acids are linked together by **peptide bonds** to form proteins. What is the source of the energy used to form these peptide bonds? ______________________________

4. Amino acid sequence in the protein is determined by base sequence in the mRNA (which was initially determined by DNA base sequence).

5. A model is available in the lab to help visualize events as mRNA is formed on the DNA and as amino acids are linked to form proteins.

6. When all mRNA codons have been read, the finished protein is released and ribosomes and tRNA's are reused. The mRNA may be used one or several times before it is broken down and its components used to form new nucleic acids.

Protein Synthesis Review — Paper Models

1. Cut apart the components for protein synthesis. (Models are found on pages 135 and 137).
2. Compare base sequence of mRNA-1 to the DNA.
3. Take mRNA-1 and begin translation at the end marked with an arrow.
4. Attach AA-1 and its tRNA to this mRNA strand.
5. Attach AA-2 and its tRNA to the mRNA strand.
6. Bond AA-1 to AA-2 and return tRNA-1 to the cytoplasmic pool for future use. Remember that amino acids bond together by **peptide linkages.**
7. Attach AA-3 and its tRNA to the mRNA strand.
8. Bond AA-2: AA-1 to AA-3 and return tRNA-3 to the cytoplasmic pool.
9. Continue until all four amino acids are linked and all tRNA's are released. The finished protein segment or polypeptide chain is then released from the ribosome.

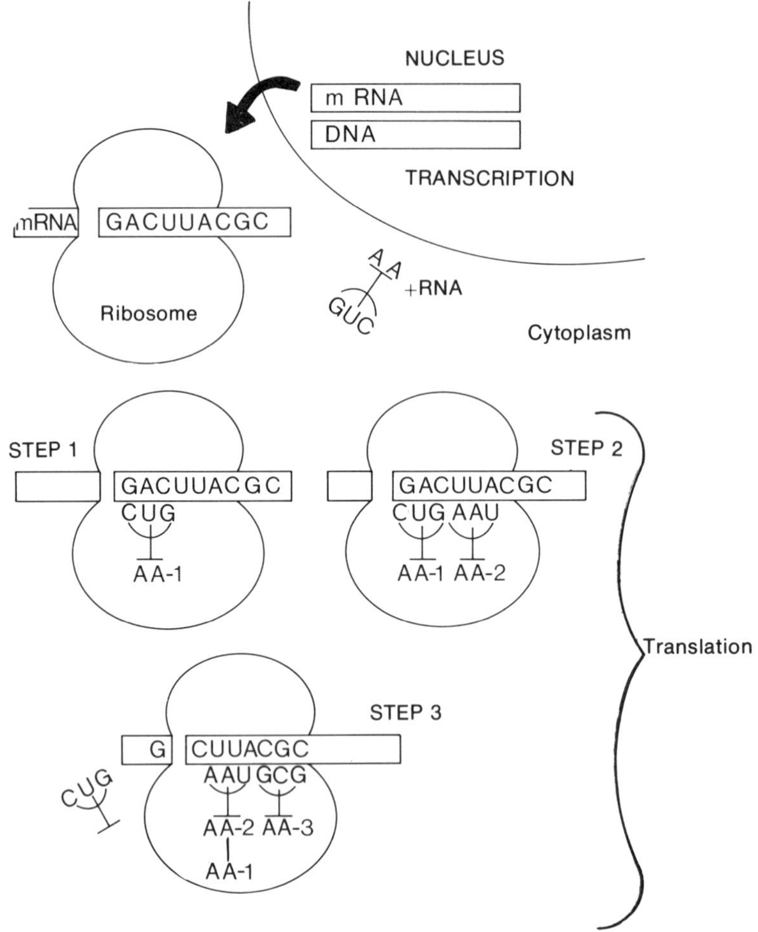

Figure 14.3. Protein synthesis.

EFFECT OF ALTERING DNA BASE SEQUENCE

Types of Mutations

Any alteration in DNA base sequence has the potential for producing a change in the resulting protein structure. Since more than one codon may code for the same amino acid, sequence change does not necessarily produce a change in the protein. A change in base sequence is a **mutation.**

Substitution mutation — If we say that UUU codes for phenylalanine, what would happen if this sequence were altered to read UUA? Would the protein synthesized from this DNA strand information be the same?______________________ Your answer should be: probably not, unless UUA also happens to code for phenylalanine. If it does not, then another amino acid will be inserted in the protein at this location. This type of mutation is a **substitution mutation** (one nitrogen base was substituted for another).

1. Compare the base sequence of mRNA-2 to that of mRNA-1.

2. mRNA-2 differs from mRNA-1 at base number__________. (Begin counting at the end marked by the arrow).

3. What change in the DNA was necessary to produce a mRNA of this type?________________

4. Codons of mRNA-1 are AAU GCC UGC GUA.

   Codons of mRNA-2 are ____ ____ ____ ____

5. Construct a protein segment using mRNA-2 and compare it to the amino acid sequence built with mRNA-1. If you are unable to correctly match a tRNA sequence, use a tRNA and its AA labeled "non-matching" (non-m). Which AA numbers are in correct sequence?____________________.

Which are incorrect?____________________

6. The cell could probably build a protein to fit the altered specifications, since the cell has at its disposal tRNA's for all codons which code for amino acids. If the amino acids hooked together were different, however, the resulting protein would also be different.

7. What other ways do you suppose the sequence of DNA bases might be altered to produce a mutation? ______________________________

Insertion Mutation — What would happen if a single base were added to the chain somewhere in the sequence?

1. Compare the base sequence of mRNA-3 with mRNA-1.

2. mRNA-3 differs at base numbers ______________________

3. Can you shift mRNA-3 to the right or to the left and get bases to match? ____________

4. What change in the DNA was necessary to produce a mRNA of this type? ____________

Codons of mRNA-3 are ___ ___ ___ ___ .

5. Construct a protein using mRNA-3 and compare it to the protein built with mRNA-1. Again use non-m tRNA and AA where necessary. Which AA are in correct order? ____________

Which are incorrect? ______________

Again, the cell would probably have been able to build the protein, but its sequence would not have been correct.

Deletion Mutation — What would happen if a single base were deleted from some place in the strand?

1. Compare base sequence of mRNA-4 to that of mRNA-1.

2. mRNA-4 differs at base numbers ______________. What change in the DNA was necessary to produce this mRNA? ______________

Codons of mRNA-4 are ___ ___ ___ ___ .

3. Construct this protein. Which AA are in correct order? ______________

Which are incorrect? ______________

Speculate on the possible effects of each of these types of alterations or mutations on the finished protein structure.

EXAMPLES OF MUTATIONS

From the previous sections it should be apparent that some types of mutations could produce drastic changes in proteins produced, with all codons beyond the mutation incorrect. Other types of mutations may alter or change only one or two codons. The latter type of mutation may or may not produce a visible or detectable change in the organism so affected.

Sickle Cell Anemia

A change in only one amino acid out of nearly 300 in the hemoglobin molecule (substitution of a valine for a glutamic acid) results in a drastic reduction in its ability to carry oxygen. This genetic defect is the basis of sickle cell anemia. Individuals homozygous for this recessive trait have blood cells of abnormal shape. Observe demonstration slides showing normal blood and blood from an individual with sickle cell anemia. Sketch a normal and some abnormal red blood cells.

Gene Control of Starch Synthesis (in peas)

The statement has been made that genes ultimately control protein structure. But how do genes influence cellular nonprotein substances such as carbohydrates and fats? Nonprotein substances are controlled and regulated by enzyme activity. What type of biological molecule are enzymes?

____

1. Let's look at an example of gene control of starch synthesis. Some peas are round and some are wrinkled. Specific genes control the production of starch from glucose by controlling the presence or absence of certain enzymes. As enzyme production is under genetic control, differences in starch content reflect genetic differences in the peas.

2. Obtain one round and one wrinkled pea (which have been soaked overnight) from the demonstration table. Cut the smooth pea in half and suspend some of the contents in a drop of water at one end of a slide. Mark that end to identify contents. Repeat at the other end of the slide using the wrinkled pea.

3. Examine these drops microscopically, comparing the starch grains. Sketch the appearance and relative size relationships of the grains in the two pea types.

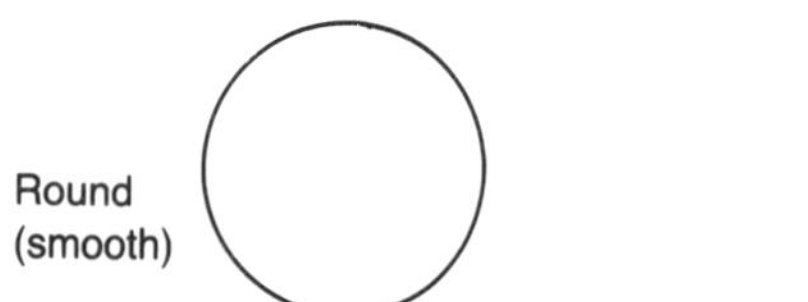

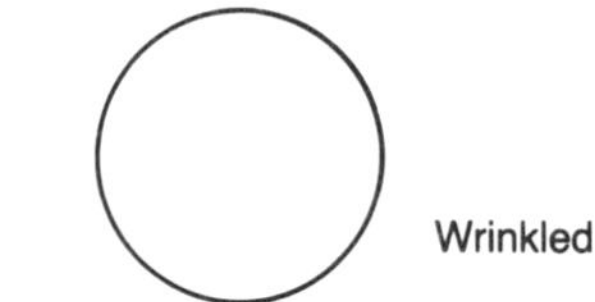

4. You will note that these peas differ phenotypically. You will also note that the starch grains differ in appearance, a reflection of biochemical differences between the two types of peas. In your text find the structural formulas of starch and glucose. What do you notice about the formulas that is the

same? ______________________________

If you can't find the answer, see Exercise 2. Enzymes convert glucose to glucose-1- phosphate (glucose +Ⓟ⟶ glucose-1-Ⓟ) and then to starch. A summary of this reaction would be:

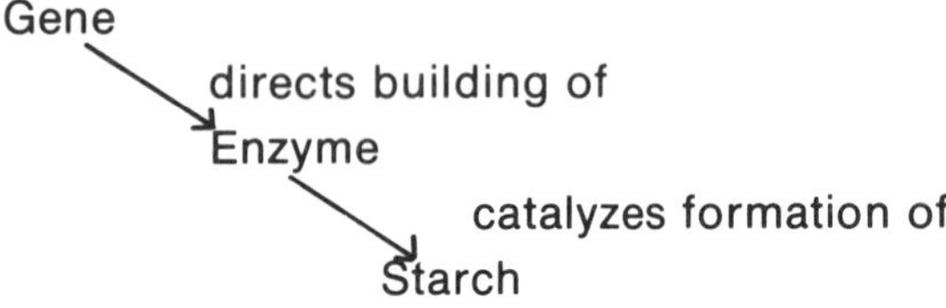

5. The information just obtained might be combined with a statement reflecting the genetics involved. On the basis of this experimental information complete the following. In homozygous wrinkled peas an enzyme concerned with starch synthesis is probably (deficient, increased). Or, we

might let R equal the gene for round and state that the gene_____gives biochemical competence for

starch synthesis which the gene___is unable to provide. At this stage we might also redefine the **gene** as: **the hereditary unit that occupies a fixed chromosomal locus, which through transcription has a special effect upon phenotype and which can mutate to various allele forms.**

Gene Control of Starch Synthesis (in corn)

1. Several ears of corn are available which demonstrate a 9:3:3:1 ratio of inheritance. This corn is from the F_2 generation of a cross of purple smooth (PPRR) with yellow rough (pprr). If you counted the four possible phenotypic combinations you would come fairly close to the expected ratio. The wrinkled or rough have less starch, just as in the peas.

Other Examples of Phenotypic Expression of Mutation

List the other examples on demonstration.

SUMMARY AND QUESTIONS

1. A nucleotide consists of ______________________________

____

2. Any base may follow another base along a single strand of DNA, but only those bases which are ______________ may be attached to form a second strand and the ladder-like configuration.

3.

	Bases		Sugar
DNA contains phosphate plus	______________	plus	________
RNA contains phosphate plus	______________	plus	________

4. In DNA, A always bonds with _____ and G with ________.

5. In RNA, A always bonds with _____ and G with ________.

6. A codon consists of

7. The process of transcription occurs in the ______________ and consists of

____

8. Translation occurs in the ______________________________ and consists of ______________________________

____

9. A ______________________________ is an alteration in DNA base sequence.

10. Three types of mutations in DNA are ______________, ______________ and ______________.

11. Indicate the type of chemical bond which links the two units.

A to T in DNA:

Amino acid to a tRNA:

Amino acid to amino acid in a protein chain:

Two single DNA strands:

12. In the space below, give the base sequence needed to form the requested molecule if the DNA strand is GATCGT. Begin reading at the left of the sequence.

Complementary strand of mRNA ______________________________

tRNA's which would attach to the mRNA ______________________________

13. In a ______________ mutation all codons except the one with the mutation are correct.

14. In a ______________ and a ______________ mutation all codons after the mutation are incorrect.

15. Enzymes are protein molecules. Why is it possible for a change in a single amino acid to completely block this enzyme's ability to function?

16. What would be the effect on a protein chain if both an insertion and a deletion mutation occurred close to each other in the DNA chain?

17. If mutations occur in a population over thousands of generations, what effects might this have on this population?

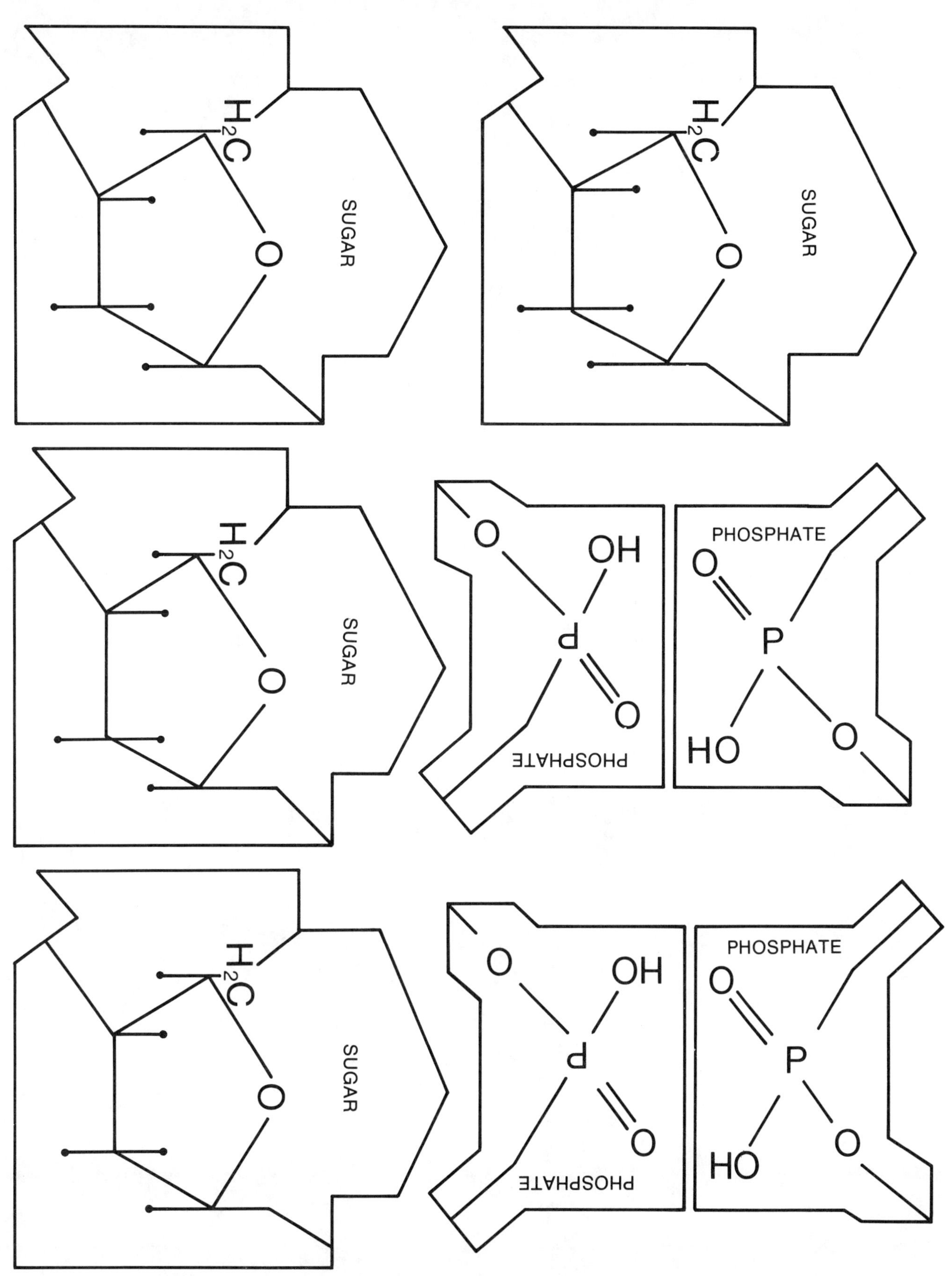

DEXOYRIBOSE AND PHOSPHATE UNITS

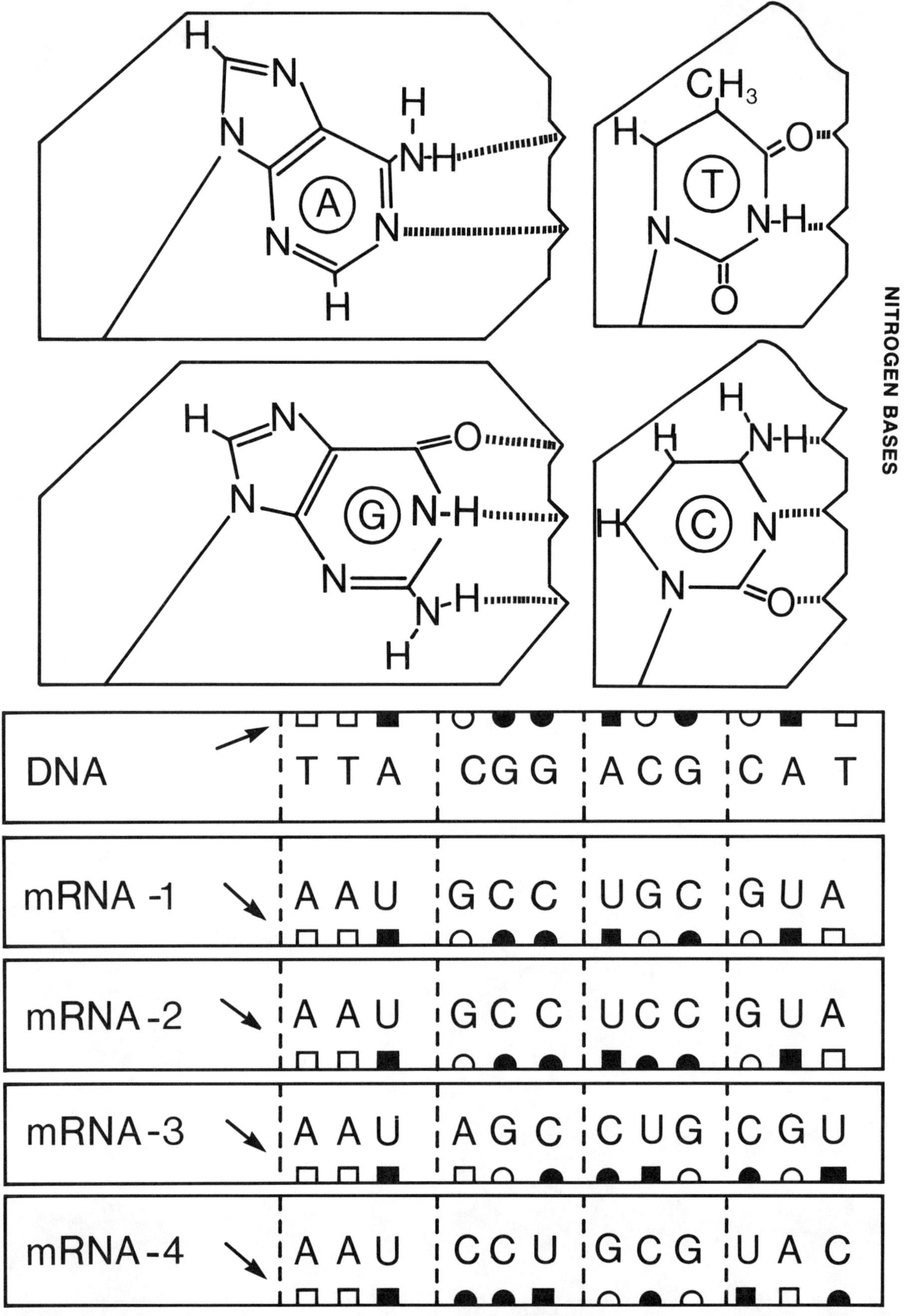

Do not cut on dotted lines. Cut into five strips.

NITROGEN BASE MODELS AND DNA AND mRNA MODELS FOR PROTEIN SYNTHESIS.

t RNA-1	t RNA-2	t RNA-3	t RNA-4	NON-M t RNA	NON-M t RNA
A A-1	A A-2	A A-3	A A-4	NON-M A A	NON-M A A
NON-M tRNA	NON-M tRNA	NON-M A A	NON-M A A		

Cut Apart to Form 16 Squares.

AA and t RNA MODELS FOR PROTEIN SYNTHESIS

LABORATORY 15
ANIMAL AND PLANT DEVELOPMENT

OBJECTIVES: 1. To study and compare the process of development in four animals: starfish, frog, chicken and mammal.
2. To study plant development.

SECTION I: ANIMAL DEVELOPMENT

The basic pattern of the developmental sequence is the same in the starfish, the frog, the chicken and the mammals. Details of development will, however, vary (and sometimes considerably) from type to type. It is possible to interpret most of these variations in an **adaptive** sense. (See text for definition of adaptation.) The organisms you will study represent four different ways of life, and it is not too surprising to see differences in reproductive and developmental machinery (as well as in other aspects of their structure and function). The starfish and frog live in the water and the chick and most mammals on land. Time intervals between fertilization and hatching for the four organisms are 1.5, 12, 21 and 281 days, respectively. Can you think of possible explanations for such time differences?

This exercise will attempt to (1) illustrate the basic patterns of development common to certain groups of animals, and (2) look at those aspects of development unique to each group and attempt to determine adaptive significance of these features.

Development can be divided into four major areas.

1. **cleavage:** subdivision of the egg cell into many cells.
2. **gastrulation:** redistribution of cells resulting in establishment of a three-layered embryonic body plan.
3. **organogenesis:** elaboration of simple embryonic structures into characteristic adult structures. The process includes
   a. **morphogenesis:** a reshaping of the embryo
   b. **cell differentiation:** structural and functional specialization
4. **growth:** increase in size

Additional terminology will be introduced as you study the various prepared slides, charts and models in the laboratory.

STARFISH DEVELOPMENT

All stages of early starfish development are included on the slide provided. Obtain one of these slides. As this is a thick preparation, use **low power only.** Starfish gametes are shed into the aquatic environment and shortly after fertilization, the zygote begins the developmental sequence. Two characteristics of this egg make it ideal for study — the egg possesses little yolk and the embryo is enclosed in a transparent membrane.

The amount and distribution of **yolk** (reserve food material) within the egg affects the way in which the zygote divides or cleaves. If there is little yolk, evenly distributed, the entire mass of the zygote is subdivided by the cell divisions and cleavage is termed **holoblastic,** as in the starfish.

Locate the various stages described in the following paragraphs and **sketch** these structures in the margin and **label.**

Unfertilized Egg

The unfertilized egg is a nearly homogeneous spherical cell with a distinct, lightly stained nucleus containing a darker nucleolus. An extracellular membrane (the **vitelline membrane**) surrounds the egg, but it adheres tightly to the plasma membrane and is not usually distinct. Based on the uniformity of staining and the size, do you estimate much yolk to be present?

Zygote

When fertilization occurs, the vitelline membrane lifts away from the plasma membrane and becomes the **fertilization membrane.** It is not always apparent on prepared sections. Since the fertilization membrane forms immediately after penetration of the cell by the sperm, can you think of a possible function for this membrane?____

The nucleus is much less conspicuous in the zygote than in the egg. Estimate the diameter in millimeters of several of these eggs and zygotes so as to obtain an approximate average. (The diameter of the low power field is 1.4 mm).

Est. Size: Egg _______ _______ _______ _______ Ave. _______

Zygote _______ _______ _______ _______ Ave. _______

2-, 4- and 8-Cell Stages

Locate a 2-cell stage. The plane of the first division is usually vertical, passing through the axis of the egg and at right angles to the equatorial plane. This first cleavage furrow marks the division between the two cells or **blastomeres.** Estimate the size of the 2-cell stage and compare it to that of the egg.

Estimated size____________, which is (smaller, larger, about the same) as the egg.

Locate a 4-cell stage. The second cleavage furrow is at right angles to the first. The vitelline membrane is usually lost by this time.

Locate an 8-cell stage. The third cleavage, which produced this stage, results in two groups of four cells, one group on top of the other. A number of cell divisions have taken place to reach this stage.

Were these divisions mitotic or meiotic?___________ Does the total mass of living matter seem to increase during this cleavage process? __________________________

Are all cells the same size?____________________

Morula and Blastula

After the 8-cell stage it becomes increasingly difficult to determine the number of cells. By about the 32-cell stage, the embryo appears as a solid ball of cells resembling a mulberry **(morula stage).**

As cleavage continues, cells become arranged around a hollow center **(blastocoel)** and the embryo is now called a **blastula.** How have cell size and numbers changed since earlier stages?

____ How can you tell that the blastula is hollow? ____

How does the diameter of a blastula compare with that of an unfertilized egg?__________________

___________________________ How many cells thick is its wall and is the wall of uniform thickness?

____

Locate the thickest side — the site where changes leading to formation of the next stage will occur. The thicker-walled half of the blastula is the **vegetal hemisphere.** These cells contain a greater amount of yolk than do cells of the opposite side (the **animal hemisphere).**

Gastrula Stage

Approximately 1.5 days after fertilization, the vegetal pole becomes flattened and collapses into the blastocoel. The first appearance of this **invagination** marks the beginning of gastrulation, and the resulting embryo is called a **gastrula.**

Two cell layers are now present, the outer **ectoderm** and the inner **endoderm.** A third layer of cells, the **mesoderm,** will soon develop as an outgrowth from the endoderm and will be located between the other two layers. These three **germ layers** are the structural building blocks which ultimately form all tissues and organs of the adult starfish.

Locate a gastrula. The pocket produced by invagination of the blastoderm is the **archenteron** or primitive gut. The opening into the archenteron (the **blastopore)** is converted into the anus. The mouth develops after gastrulation when the blind end of the gut perforates through the ectoderm to produce a new opening. The embryo now possesses a complete alimentary tract with openings at either end. After a day or two, the gastrula is converted into a free-swimming larva which will eventually settle to the bottom and **metamorphose** into an adult starfish.

Three-dimensional models of the various stages are available for further study after you have completed the sketches.

FROG DEVELOPMENT

Amphibian eggs contain a larger quantity of yolk than the starfish, and yolk is unevenly distributed. The type of cleavage can still be described as holoblastic, but the presence of yolk at one side will lead to differences in cell size. In frogs the first cleavage occurs approximately 90 minutes after fertilization and later cleavages occur every 20-30 minutes. Place drawings of the cleavage phases in the margin.

Cleavage

Slides are available showing both early and late cleavage stages. Are all cells the same size?__________. In what way, if any, does the 2-cell or the 4-cell stage differ from the same stage in the starfish? ____

A thin layer of dark brown pigment just beneath the surface characterizes the animal hemisphere of the cleaving zygote. What substance probably causes the lighter appearance of the vegetal pole?

____

The vegetal hemisphere lacks pigment but is packed with large **yolk platelets.** Beginning with development of the second cleavage plane, cells of the animal hemisphere tend to divide faster, leading to smaller cells than in the vegetal hemisphere. A typical morula stage does not occur.

Blastula

Study a slide of a frog blastula. Is a blastocoele seen?_____ At which pole is it located?_______

How many cell layers are in the upper wall of the blastocoele?_________ Do you think this blastula could invaginate? _________

Cells of the blastula are apparently **undifferentiated,** differing only in size. Cells of the animal region continue to divide more rapidly than vegetal cells. Is the frog blastula hollow like the starfish?_________ How has the darkly pigmented area changed since the earlier stages?________

____

Label **Figure 15.1A.**

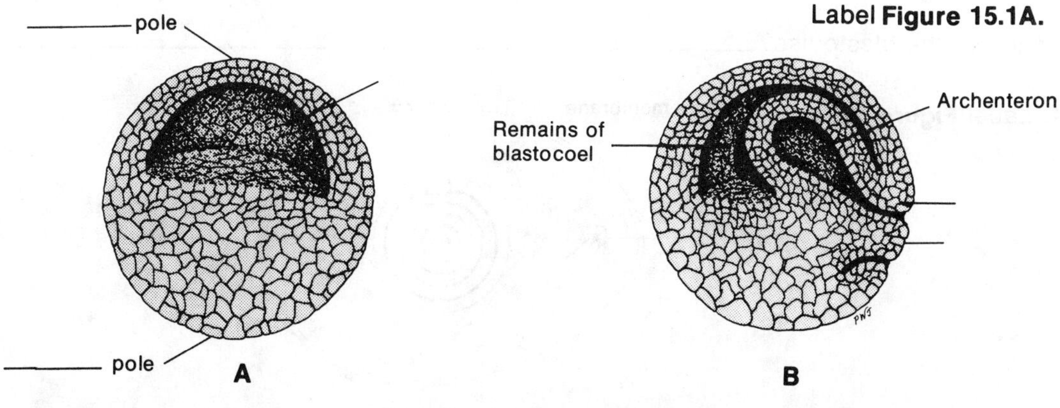

Figure 15.1. Frog development. **A,** cross section of a blastula; **B,** cross section of a gastrula.

Gastrula (Demonstration)

After the blastula stage, differences between echinoderm and amphibian development are even more marked. The presence of the large quantity of yolk produces changes in appearance and location of archenteron, blastopore and the three germ layers, although the basic pattern is the same. Pigmented cells have moved down in a sheet (much like syrup flowing down over a ball) and the entire embryo surface is pigmented except for a small, sharply defined yellow area **(yolk plug).** Embryos in this stage are undergoing gastrulation. It is considerably more complicated than in the starfish. Could a simple invagination occur in this area?______It is necessary for cells around the rim of the yolk plug to move in and under, producing the endodermal and ectodermal layers. The area where these cells move in is called the **dorsal lip** of the blastopore or the **primary organizer** (because it induces nearby cells to change into distinct cell types). The yolk plug will eventually disappear and the embryo will elongate.

Observe the demonstration slide and models and label **Figure 15.1B.**

CHICK EMBRYO DEVELOPMENT

Both developmental series studied take place in eggs deposited in an aqueous environment. Before animals can become terrestrial, a different type of egg (not susceptible to desiccation) had to develop. Reptile and bird eggs contain a series of extraembryonic membranes (simulating the watery environment) and a water-proof shell to prevent drying. The membranes are:

amnion: fluid-filled sac surrounding and cushioning the embryo
yolk sac: contains nutritional material
allantois: stores wastes of the embryo; its blood vessels lie near the shell and function in gas exchange.
chorion: outer membrane surrounding embryo and other membranes.

The egg, consisting largely of a mass of yolk, is fertilized in the oviduct. Cleavage begins immediately following fertilization, and by the time the egg has its white (albumen) and shell, the embryo has reached a modified blastula stage or blastoderm. Further development will proceed only when incubation at 37.5 C (approximately 99 F) is started. Complete development will require 21 days of incubation.

The Unfertilized Egg (work in groups of four)

1. Carefully break a raw egg and empty the contents into a dish of water. On the shell note the **membranes** and the **air space.** Would the shell and membranes be permeable to gases?______to water?________

2. Use a probe to note the density of the albumen. Is it of uniform thickness and density?____________You should note two spirally wound strands **(chalazae)** which are more dense. One is found in the blunt end, and one at the more pointed end of the egg. What is the function of these strands?

3. Examine the top of the yolk for a small white spot **(blastodisc).** If yolk is food, what is the function of the blastodisc?____

4. Label **Figure 15.2.**

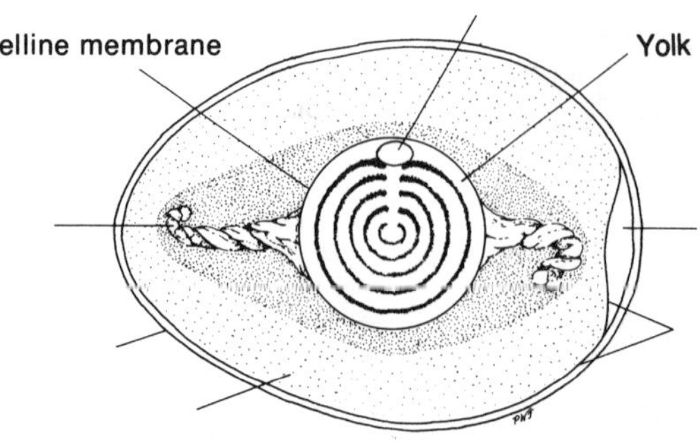

Figure 15.2. An unfertilized bird egg.

Cleavage and Gastrulation

Cleavage in bird eggs begins as it does in frog eggs, but the large mass of inert yolk is so great that the yolk material is not divided **(meroblastic cleavage).** Cleavage is limited to the small disc of cytoplasm on the yolk surface at the animal pole and leads to a disc-shaped mass several cells thick floating on the yolk surface. This mass of cells, the **blastoderm,** is a modified form of the blastula seen in starfish and frog embryos, but the large quantity of yolk prevents the formation of an open blastopore. The **primitive streak** is homologous to the blastopore, and cellular migration in this region is similar to that around the lip of the open blastopore in other organisms.

24-Hour Embryo (Demonstration)

Note the segments **(somites).** There are probably three or four pairs. These structures also occur in mammals. Is segmentation found only in animals with vertebrae or bony backbones? ______________

Can you think of some segmented animals? ____

The shallow fold at one end of the embryo marks the head region. The thin line running the length of the embryo is called the **notochord.** This is a flexible support rod and in later exercises we will look at its significance from an evolutionary and classification standpoint.

33-Hour Embryo (Demonstration)

By this time the embryo contains a well-defined brain. The prominent bulge at the anterior end is the forebrain, and the two lesser swellings behind it are the mid- and hind-brain. The heart is a bent tube located just posteriorly to the forebrain. How many somites do you see?__________ Somites will later give rise to muscle, some connective tissue and the vertebral column.

48-Hour Embryo (Demonstration)

Various organs are now formed. Locate as many of these as you can. In the anterior (head) region note the **gill slits.** Does it seem strange that a chick embryo has gill slits?________________

Do adult chickens have gill slits?__________ Are somites still visible at this time? ____________

MAMMALIAN DEVELOPMENT

Most mammals develop within the uterus of the female. The general sequence of development is much like that of bird embryos, but extraembryonic membranes have become adapted to placentation. The embryo is attached to the wall of the uterus by the **placenta,** which forms partly from the uterus. Nutrients and respiratory gases are received from the mother through the placenta. If this is the case, would you expect a mammalian egg to contain a large or a small amount of yolk?______

Mammalian Ovum (Demonstration)

On low power observe the relatively large space with the structure projecting into it. In the mature follicle the ovum is visible inside this projection. A very distinct wall or layer surrounds the ovum. The nucleus is distinct and is surrounded by a spongy appearing material. Switch over to high power and observe the ovum structure. **Figure 15.3** relates the events of meiosis to what you see in the mammalian ovary.

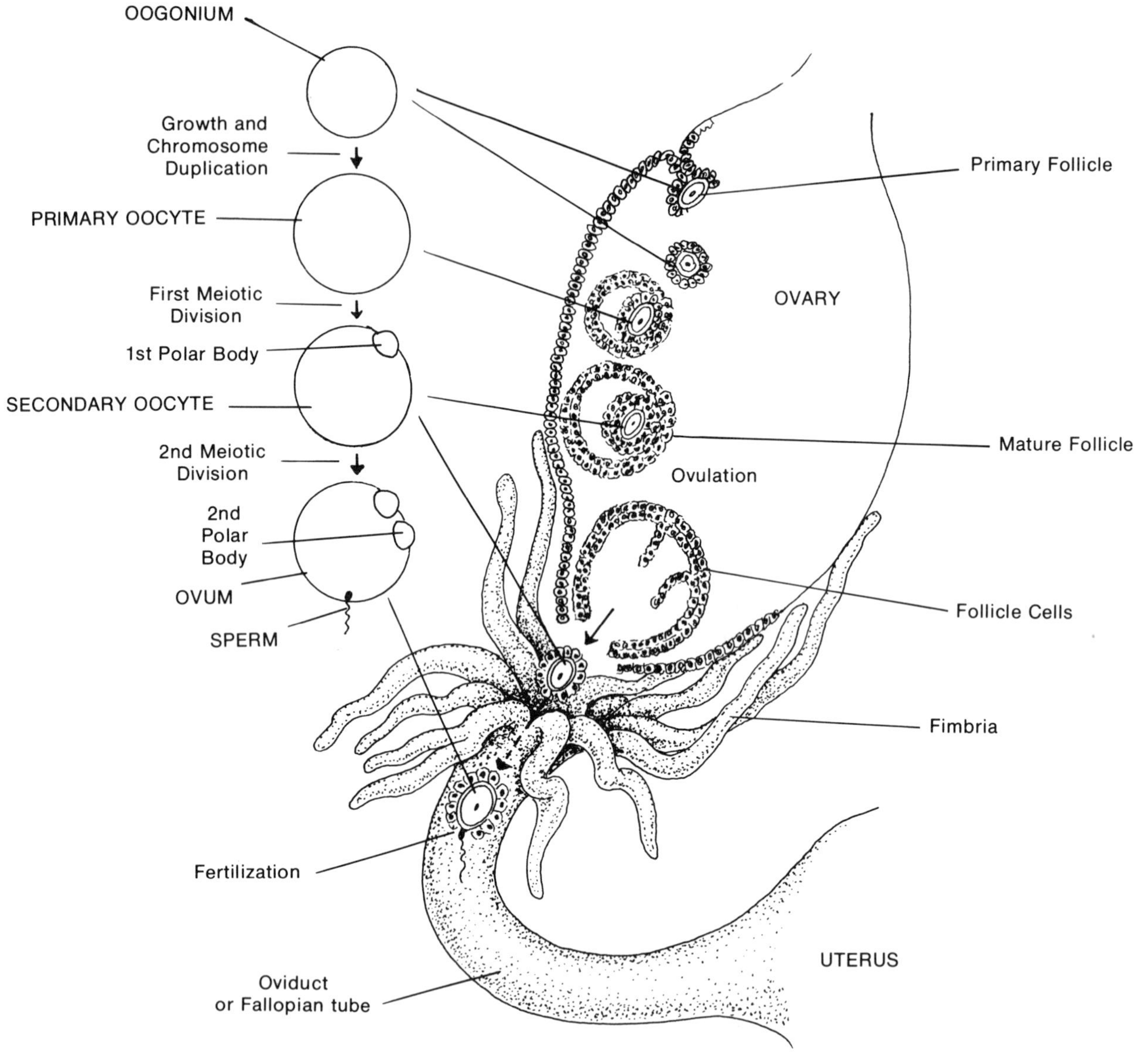

Figure 15.3. Oogenesis in the mammalian ovary.

Later Development

1. Study the drawings in **Figure 15.4.** If you were handed one of these embryos, could you really determine the type of embryo you held?______What similarities are noted?____________________

____

What differences?____

2. Study the 10 mm pig embryo on demonstration. This is a mammalian embryo at a stage about equivalent to six weeks gestation in the human. Prominent structures include the brain, eyes, heart, liver, kidney (embryonic) and limbs. Find the somites. Note the umbilical cord with its blood vessels.

3. Also observe the human fetuses and charts on display.

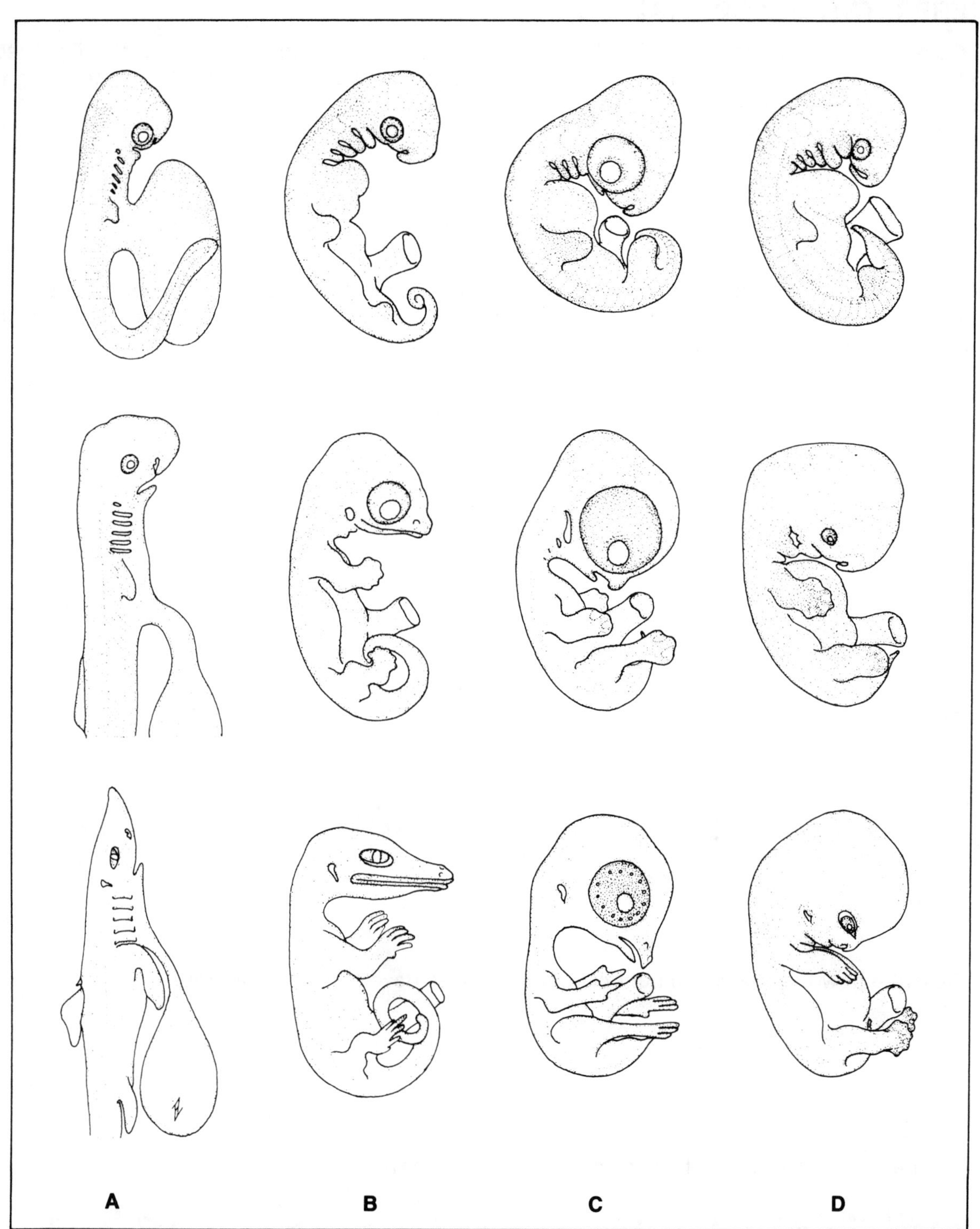

Figure 15.4. Comparison of vertebrate embryos from different classes. **A,** Shark; **B,** Lizard; **C,** Chick; **D,** Human. (From H.E. Lehman, *Chordate Development.* Winston-Salem: Hunter Publishing Company, 1977).

INTERPRETATION AND ANALYSIS

1. Compare your observations on chick and mammal development, and compare both with those on starfish and frog. What developmental features of the land vertebrates would you feel might represent specific adaptations to a terrestrial life?

2. What are the advantages and disadvantages of development by means of (a) eggs enclosed in shells and (b) placental mechanisms? What properties would be essential in an eggshell which serves as an enclosure for an embryo developing on dry land?

3. List some of the adult tissues the three embryonic germ layers will produce. How do adult tissues arise from a germ layer?

4. Describe how the relative amount of yolk influences developmental patterns.

5. Differentiate between holoblastic and meroblastic cleavage.

6. You are given a slide and told only that it represents development of an amphibian, a bird or a starfish. How would you proceed to identify your slide?

7. Compare starfish and amphibian development.

	Starfish	**Amphibian**
Cleavage		
Blastula		
Animal Hemisphere		
Vegetal Hemisphere		

Building a Developmental Vocabulary

______________________	The hollow ball of cells stage seen during starfish and frog development.
______________________	The embryonic germ layer which gives rise to brain and spinal cord.
______________________	A dorsal, rodlike structure which supports the vertebrate embryo.
______________________	A saclike structure functioning in excretion (found in the chick embryo).
______________________	A cavity (formed in the gastrula stage) destined to become the gut cavity.
______________________	The lighter (nonpigmented) hemisphere of the frog egg.
______________________	The repeated division of the zygote.
______________________	A membrane (surrounding the chick embryo) which contains fluid for protection.
______________________	The embryonic germ layer which gives rise to bone and muscle.
______________________	The area of the chick egg, located on the yolk, where the embryo develops.
______________________	The invagination process which forms the gut cavity of the frog and starfish.
______________________	String-like structures in the chicken egg which help to maintain the embryo in an upright position regardless of how egg is rotated.

SECTION II: PLANT DEVELOPMENT

Seeds

Higher plants reproduce by means of a highly evolved unit called the seed. A **seed** contains the embryo plant, stored food for early development of the seedling, and a protective coating. The seed results from the fertilization of the ovule by a pollen grain on the parent plant.

In the bean seed, the bulk of the seed consists of the two **cotyledons,** the seed leaves. Cotyledons are rich in stored foods — proteins, oils and starch — and are utilized by the embryo in its early rapid growth. Man has made use of these storehouses of food in his agriculture. For example, the soybean seed is an important source of oils for cooking oils and margarine, paint bases, etc. The proteins are utilized as feed supplements for livestock and are now eaten directly by man in such delicacies as bacon bits and soy hamburgers.

The **embryo** is found between the two cotyledons, off to one side of the seed. The embryo contains a rudimentary shoot (the **epicotyl)** and root **(hypocotyl).** The terms hypocotyl and epicotyl refer to position relative to the cotyledon attachment. "Epi-" means above and "hypo" means below.

1. Examine a bean seedling which has been soaked in water to soften the protective coating. Carefully split it apart between the two cotyledons. Look for the embryo. You may need a hand lens. The epicotyl may be identified by the presence of immature leaves. Compare what you see to **Figure 15.5.**
2. Label **embryo, cotyledon, seed coat, epicotyl** and **hypocotyl.**

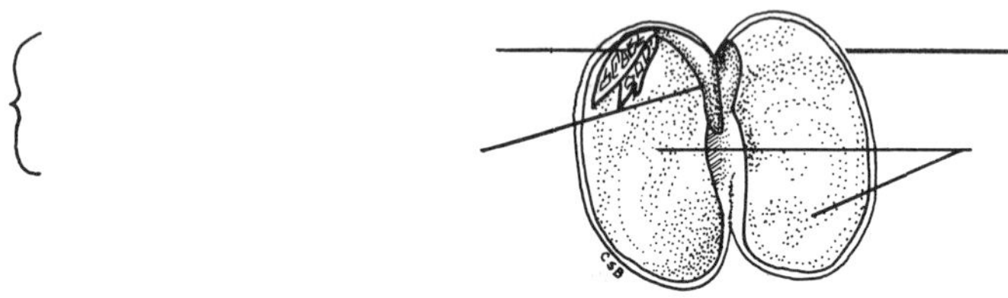

Figure 15.5. Bean seed.

Seedling

During its early development, a young plant is called a **seedling.** The early growth pattern of a seedling is critical to its survival. First to emerge from the seed is the root. The root grows rapidly, its cells dividing many times. A well developed root system is important in supplying water and inorganic nutrients to the seedling as well as anchoring it.

The developing shoot is fragile and is usually protected as it grows to the soil surface. For example, in the bean, the tip of the shoot does not emerge first from the soil. In fact, the shoot tip is pulled from the soil by a hook in the stem below the cotyledons. This hook is tough and is able to push its way through the soil. Once the shoot emerges from the soil, the hook straightens out and the shoot tip moves into a superior position.

1. Bean seedlings (living and in plastic blocks) in various stages of development are available. Make a series of four sketches showing the time sequence of the development of the bean seedling. Label one drawing.

2. Seedling structures correspond to features of the seed which have developed. List the seed part which developed into each seedling part.

_________________ leaves _________________ hook

_________________ seed leaves _________________ stem

_________________ root system

3. What happens to the size and appearance of the cotyledons as the seedling matures?

____

4. Beans belong to a group of plants referred to as **dicots,** due to the presence of two cotyledons. Corn, however, is a **monocot.** Observe the corn seedlings and note the ways monocots and dicots differ. The cotyledon supplies organic food and inorganic nutrients for the seedling. What is the source of these for the mature plant? _______________________________

Growth of the Plant

Growth in the plant results from formation of new cells and enlargement of these cells. Cell division is localized in areas called **meristems.** Meristems are found at the tip of the shoot and at each root tip.

A growing root is divided into four fairly distinct regions. At the tip of the root is the **root cap,** a protective region which is pushed through the soil as the root grows. Behind the root cap is the **meristem** where cell division occurs at a rapid rate. New cells are produced here — some replace worn off root cap cells and others become cells of the main part of the root. Behind the meristem is the **zone of elongation** where newly formed cells grow larger. As they grow, they push the meristem and the root cap through the soil. Behind the region of elongation is the **zone of differentiation.** In this area, cells are transformed into different functional units. For example, some cells become xylem cells and function in water transport; others, on the surface, become epidermal cells with root hairs and function in water absorption.

1. Examine a slide of a longitudinal section of an onion root tip. These sections have been stained to show the nuclei distinctly. In **Figure 15.6** label the four regions of the root tip.

2. In the shoot, the meristem is at the very tip. Why would this arrangement not be successful in root tips? _______________________________

3. Which region would be best to use in an exercise on mitosis? _________________

Why? _______________________________

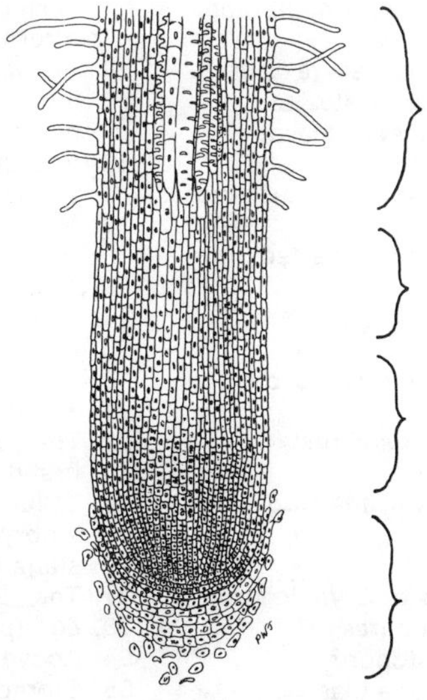

Figure 15.6. Longitudinal section of an onion root tip.

ANIMAL AND PLANT DEVELOPMENT CROSSWORD PUZZLE

Across

1. A frog blastula is dark on one side because of brown____.
3. structure which keeps embryo upright when bird egg is rotated
5. A chicken egg has an____space near the large end.
6. The outer tissue layer is the____derm.
8. The third tissue layer to form is the____derm.
11. Unfertilized____are surrounded by a vitelline membrane.
12. the fertilized egg
13. zone or area separating 2 cells in the 2-cell stage (plural)
15. The small white spot on the yolk is the blasto____.
17. Frog and starfish eggs undergo____blastic cleavage.
18. The root cap protects the____stem.
20. In 24 and 33-hr chick____one can see somites and notochords.
24. In the starfish the blastopore will be converted into the____.
25. A solid ball of cells is a____ula.
27. A blasto____is the cavity in a blastula.
28. structure which allowed reproductive success on dry land
30. part of bean stem which emerges from soil first
31. The____pole contains yolk platelets.
32. ____vagination of the blastula occurs in the frog blastula.
34. The rudimentary shoot is the____yl.
36. ____genesis: the elaboration of simple embryonic structures to form characteristic adult structures.
37. the organizer or dorsal____of the blastopore
38. Cleavage stops with the formation of the blas____.
39. There are 3 tissue or____layers in the organisms you studied.
40. leads to production of many similar, but smaller cells
41. develops from the epicotyl

Down

1. The opening into the archenteron is the blasto____.
2. The support rod in the chick and pig embryo is a____chord.
3. Cleavage results in many identical____.
4. first 3 letters of the membrane involved in waste product storage in chick embryos
5. first 3 letters of the membrane which is fluid-filled and cushions the embryo.
7. After organogenesis, all the embryo does is____.
8. Reshaping of the embryo is called____genesis.
9. In the bean plant the____tip is pulled from the soil by the hook.
10. In____blastic cleavage, yolk is not divided.
11. The second tissue layer formed is the____derm.
12. Mitosis occurs in the____of multiplication.
14. Higher plants reproduce by means of____.
16. a segment visible in the chick and pig embryo
17. The root system develops from the____cotyl.
18. Inside the egg shell are a number of____.
19. At the tip of the____is a cap.
21. The quantity of____greatly influences gastrulation.
22. The yolk is contained in____. (plural)
23. first 6 letters of the primitive gut
25. cellular process which forms a blastula
26. embryos of mammals receive food from this structure
29. stage in which the archenteron is found
31. The____line membrane disappears after fertilization.
33. egg (plural)
34. Above the zone of multiplication is the zone of____ation.
35. Stored food in a bean is in the two____edons.

(See page 208 for correct answers.)

LABORATORY 16
ADAPTATION AND EVOLUTION

Looking back over the previous semester and the first part of this semester you should be aware of two general ideas. In coping with their environment, all organisms face similar problems. They have the problems of assimilating raw materials, excreting wastes and transporting substances internally. There are also the problems of support, coordination and reproduction. They must be able to conduct these necessities of life in the particular environment in which they are found. The mechanisms utilized by an organism to cope with the particular problems of its environment are called **adaptations.** Adaptations may be structural, behavioral or physiological. Survival of populations of organisms will depend upon the ability to change. The changes will be gradual over a long period of time, since populations evolve, not individuals. Genetic differences in a population provide **natural selection** with variety of material upon which it can act. By selecting for individual organisms better able to cope with the changing environment, populations undergo continual adjustment to shifting environmental demands.

In the laboratory you will observe a variety of adaptations. **Concentrate on how these adaptations are unique to this organism and relate the adaptations to environmental pressures to which the particular organism is subjected.**

PLANT LEAF ADAPTATIONS

This section will be concerned with some morphological adaptations that allow plants to survive in a certain habitat, particularly structural changes that have evolved with respect to the ability to obtain and retain water. On the basis of available water supply, it is possible to divide plants into three major categories.

Mesophytes

Obtain a slide with all three leaf types on it. Your instructor will tell you how to locate the mesophytic leaf. This section is typical of the **mesophytes** which inhabit regions of average water conditions. Most wild and cultivated plants of the temperate regions are of this type.

1. Using **Figure 16.1** and the following description, locate on the slide the structures listed.

cuticle — waxy outer covering
epidermis — thin layer of cells
palisade cells — elongated cells containing many chloroplasts. Principal site of photosynthesis.
spongy layer — irregular cells immediately below palisade layer. Many air spaces separate these cells.
epidermis — contains frequent stomata, each with a pair of surrounding guard cells
cuticle

2. Add the appropriate labels to **Figure 16.1.**

3. What is the function of the cuticle? ______________________________

4. How many cell layers are in the epidermis? ______________________________

5. How many stomata do you see? ________ Are they located on the upper or lower surface?

____

6. How many vascular bundles do you find? ________________

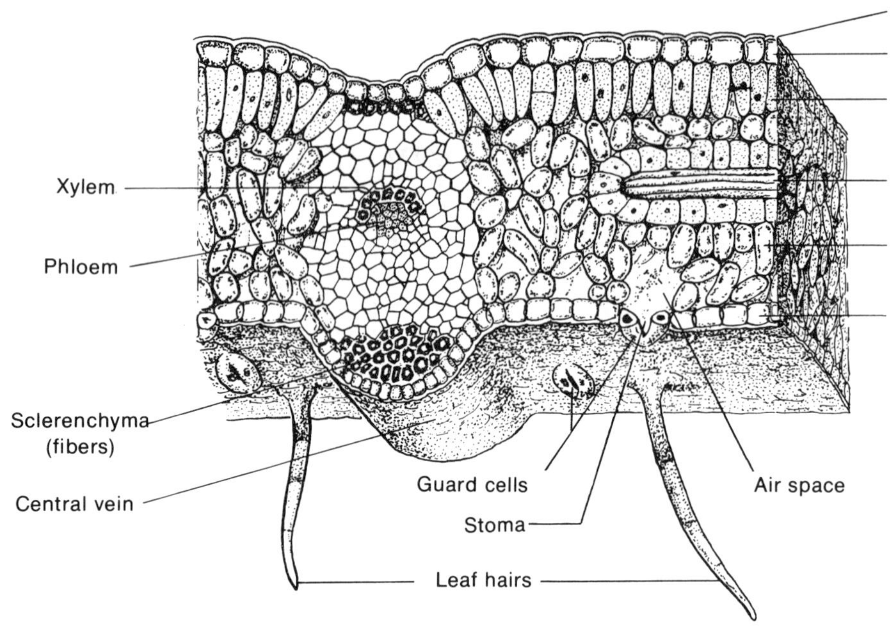

Figure 16.1. Cross section of a mesophytic, dicot leaf.

(From P.W. Johnson, *General Biology in the Laboratory*, Hunter Publishing Co., 1977.)

Xerophytes

Xerophytes include a number of species found in environments where the water supply is physically or physiologically deficient. Leaves of xerophytes have modifications that contribute to their survival during dry periods. Some of these modifications are (a) less surface area, (b) thick cuticle and epidermal walls, (c) depressed stomata, (d) more palisade tissue than spongy tissue, (e) increased conductive tissues, (f) smaller, more densely arranged cells, and (g) ability to fold or roll the leaf. All of these modifications will not occur in every xerophyte.

Observe this section on your slide and note which of the above modifications are found in the leaf of this particular plant. Sketch and label a portion of this section to show how it differs from mesophytic leaves.

Hydrophytes

Hydrophytes may float on the surface of the water or be submerged at various depths. Focus on the third leaf type on the slide.

1. The large spaces are air chambers. Why should this plant need these? ____________________

____

2. Can you find any vascular bundles? ______ Would you expect to find these increased or decreased in number (when compared to the mesophyte)? ____________________ Why? __________

____

3. How are the cuticle and epidermis modified? ______________________________
____

4. How are stomata modified? ______________________________

5. Sketch a portion of this leaf type to illustrate how it differs from the other two types.

6. Complete the chart, comparing the xerophyte and hydrophyte to the mesophyte. By thinking about the plant's habitat and utilizing your sketches, you should complete this chart at a later time as a review.

Plant Type	Air Chambers	Protective Tissues	Leaf Modifications (stomata, other)	Vascular Tissue
Mesophyte				
Xerophyte				
Hydrophyte				

INSECT ADAPTATIONS — DEMONSTRATION

Insects exhibit a large number of adaptations, many of which aid the insect in avoiding predators. These adaptations may take the form of **cryptic** or **concealing coloration, revealing coloration** (to frighten predators) or various forms of mimicry. **Batesian mimicry** is exhibited by harmless species which resemble forms that are distasteful or poisonous. **Mullerian mimicry** exists when there is a similarity in form or appearance between species which may or may not be closely related to each other. Mullerian mimics are all distasteful or poisonous and each profits by the predator's experience with a similar pattern.

Observe the various examples which are on display. Note any additional information available to you on cards located beside the various insects. Reference sketches or notes of some of these examples should be made.

ADAPTATIONS FOR PARASITIC LIVING

Examples of parasitic living may be found in both plants and animals. A **parasite** lives in or on another organism at the expense of the latter (the **host).** Much evidence exists to indicate that parasitic forms have been derived from free-living forms in the process of evolution. Generally speaking, the majority of such adaptations might be considered as retrogressive or degenerative. For example, one usually sees a reduction in nervous and locomotor organs in parasites. Would a parasite be as dependent for its survival on a well-developed nervous system as a free-living organism? Extensive changes are also seen in the reproductive systems. Many have complex life cycles and asexual reproduction, insuring the production of large numbers of offspring. The oriental liver fluke, *Clonorchis sinensis,* requires three hosts to complete its life cycle: a snail, a fish and man (or some other mammal). Can you think of a good reason for the development of such an adaptation? It must have some survival advantages. No individual host survives forever. The species could be perpetuated by being able to change hosts. Seldom is the parasite able to move directly from its present host to another host of the same species. Many parasites therefore show elaborate cycles with as many as two or three intermediate hosts and/or a free-living larval stage.

Anatomy of *C. sinensis*

1. Obtain a slide of *C. sinensis* and focus on it with **low power.** First look at the structures and systems which might serve as adaptations to better fit it for survival as a parasite. Compare the specimen to **Figure 16.2** as you look for the various organs and structures.

2. Note the flattened, leaflike shape of the worm. The digestive system in this parasite is not extensive. How might the shape and environment of the organism relate to a poorly developed or less complex digestive system?____

____

3. On the outer surface of the worm is a thick **cuticle,** highly resistant to enzyme action. Why would protection of this sort be advantageous to this parasite? ____________________________________

____

4. Observe the **anterior (oral) sucker** which has the mouth at its center. Behind the mouth is the muscular **pharynx** which helps withdraw fluid from the host and move it into the parasite's intestine. There is also a **ventral sucker** approximately one third of the way from the anterior end. Of what advantage would a sucker be to a parasite?____

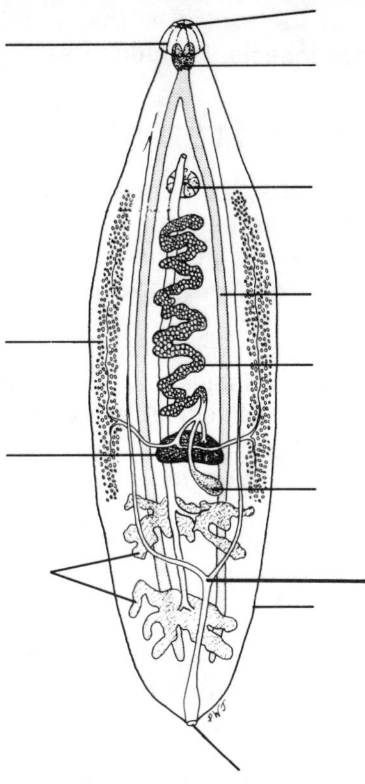

Figure 16.2. *C. sinensis* anatomy.

Labels for these and other structures may be added to **Figure 16.2** now (or later as a means of review). Primarily you should be concerned with comparing the figure to the actual organism.

5. The **intestine** continues posteriorly from the pharynx and almost immediately divides into two long sacs. There is no opening to the intestine other than the mouth.

6. In the posterior third of the body are a pair of irregularly branched **testes.**

7. The long coiled **uterus** lies behind the ventral sucker and between the intestinal lobes. It is usually filled with brown-stained embryos.

8. The single **ovary,** located near the midline, is connected to a lighter staining **seminal receptacle** (in which sperm are stored). Also connected to the ovary by two delicate tubules are the **yolk glands.** The latter consist of many small rounded bodies characteristically located laterally about the mid-portion of the body.

9. The fluke is **hermaphroditic** and capable of self-fertilization (although cross fertilization may occasionally occur). What advantage is gained from possessing this capacity for self-fertilization?

____

10. At the posterior end of the worm is the **excretory pore.** An excretory tubule leads anteriorly and at about the level of the seminal receptacle it divides into two **excretory tubules** which run anteriorly just lateral to the intestine. These are stained red on some slides and on others are light and unstained.

11. In comparing the digestive, excretory, reproductive and nervous systems — which are extensive and which are reduced in importance in this parasite? ______________________________

____

____

ADAPTATIONS DEMONSTRATED BY A STUDY OF FOSSILS

You have been studying adaptations as they exist today in different groups of organisms. What changes were necessary in the past to arrive at the diversity of forms you see today? How long did it take? What were the environmental changes which directed the route the adaptations took? To see how these questions have been answered we need to look first at the basic principles governing evolutionary change.

The concept of **evolution,** the changes occurring within a group of organisms over a long period of time, is not a simple one. Many complex variables may have interacted to produce the structural and functional changes observed. The mechanism of evolution seems to include several basic principles:

1. There is genetic continuity from generation to generation and offspring tend to resemble parents, barring mutation.

2. Variation in gene characteristics exists within an inter-breeding population. An individual may be tall or short, blonde or brunette, plaid or striped, etc. How do new variations arise?

____

3. In nature, the number of individuals in any generation which survive to reproductive age is far less than the total number of individuals born during that generation.

4. Those individuals which survive and reproduce are the ones best adapted to their environment. This is the principle of **natural selection.**

Only those organisms which produce offspring can return their genetic content to the **gene pool** to be used by the next generation. Natural selection thus directly influences the hereditary content of each generation. The direction of evolutionary change is determined by the selective pressures of the environment acting on variation present in the gene pool.

Paleontology (the study of the life of past geologic eras) is one discipline used to study evolutionary relationships. A major source of paleontologic data is the fossil record, which includes preserved parts of ancient organisms or other signs of their presence. Fossil age is estimated by dating rock strata in which they are found. There are a number of difficulties involved in determining evolutionary relationships using fossil data only. Thus, modern taxonomy also utilizes other sources — development, comparative anatomy and physiology, etc.

Perhaps the most complete fossil record belonging to any animal group is that of the horse family. This record extends with very few gaps over fifty million years through a sequence of genera. As the horse's evolutionary history is relatively complete it can be used (1) to demonstrate the effects of environmental selection pressures and (2) to illustrate the use of fossil data in determining evolutionary and taxonomic relationships. Skulls, legs and charts should be studied in lab. The graph can be completed later.

Skull Structure

The earliest genus was *Hyracotherium* (also known as *Eohippus).* The largest species were about 20 inches tall, with dog-like bodies and four toes with pad-like hoofs on each forelimb. Look at the model of the skull of the first horse and compare it to the skull of the modern horse. In addition to size, can you see other differences? Is the orbit (bony socket of the eye) located approximately midway or is there greater muzzle length in one? ____________________ Does the **relative** (not the actual) size of the brain appear to have increased or decreased? ____________________ Does the number of teeth differ? ____________________ Is the orbit different? ____________________________________

Also available is a model of a horse skull *(Miohippus).* This horse lived approximately midway between the time of *Hyracotherium* and *Equus* (the modern horse). What features does it share with the modern horse? with the first horse? ____________________________________

____

Tooth Size and Structure

In horses the cheek teeth or grinding teeth are separated from the others by a space. **Figure 16.3** illustrates their position in modern horses. The combined widths of the cheek teeth have been found to be closely correlated with the size of the jaws and skull and the overall size of the animal.

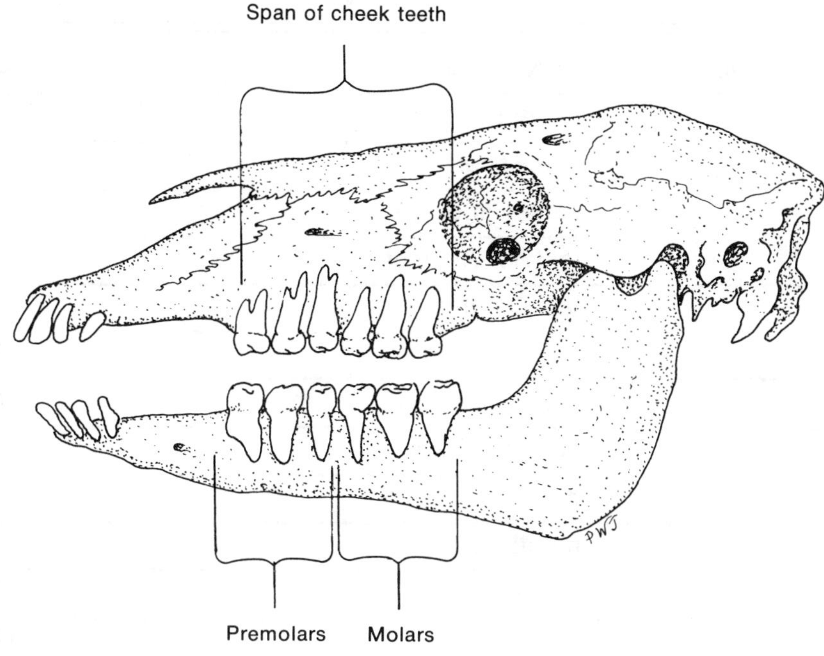

Figure 16.3. Skull of modern horse, *Equus caballus.*

1. The use of fossils will be illustrated by combining data in a graph. Detach the graph at the end of the exercise. On this graph plot the data given in Table 16.1. (P.S. Don't panic at the names. The exercise is concerned with general principles — not learning these genera.) As each point is plotted, place above it the name of the genus it represents. **Do not connect points.**

2. Next, refer to **Figure 16.4** showing stages in the evolution of teeth. You need to correlate the data in this figure and your graph. To do this, use the sequence of genera in this diagram to connect points representing corresponding genera on the graph. For example, connect the points representing *Hyracotherium* and *Epihippus* (your line should also pass through the point representing *Orohippus).* Continue in the same fashion, using all the genera on the diagram. If more than one point is given for a particular genus, connect them. The point representing the most recent data should then be used to continue the graph line. Connect the points representing *Hypohippus* and *Megahippus.*

3. Your graph now represents the currently accepted evolutionary taxonomy of these fossil genera. Many tooth characteristics in addition to span and grinding surface area have been used to establish this evolutionary scheme. These include presence or absence of a hard, durable cement between the tooth ridges, an increasingly complex ridge pattern, and an increase in both tooth length and growth period (permitting greater wear). Some of these characteristics are illustrated on charts posted in the lab.

4. What evidence can you find that the direction of an evolutionary change may be reversed? Cite such an occurrence in at least two genera and support your conclusion with data from the graph.

TABLE 16.1
SPAN OF UPPER CHEEK TEETH IN FOSSIL HORSES

GENUS	TIME OF EXISTENCE	SPAN OF UPPER CHEEK TEETH (cms.)
Hyracotherium (Eohippus)	early Eocene	4.3*
Orohippus	middle Eocene	4.3
Epihippus	late Eocene	4.7
Mesohippus	early Oligocene	7.2
	middle Oligocene	7.3
Miohippus	late Oligocene	8.4
	early Miocene	8.3
Parahippus	early Miocene	10.0**
Anchitherium	early Miocene	11.3
Archaeohippus	middle Miocene	6.5
Merychippus	middle Miocene	10.2
	late Miocene	12.5
Hypohippus	late Miocene	14.2
Megahippus	early Pliocene	21.5
Pliohippus	early Pliocene	15.5
	middle Pliocene	15.6
Nannippus	early Pliocene	11.0
	late Pliocene	10.7
Neohipparion	middle Pliocene	13.1
Equus	late Pliocene	18.8
	Pleistocene	17.6

* Plot all points in the center of each time division.
** Plot this point three-fourths of the way through the early miocene.

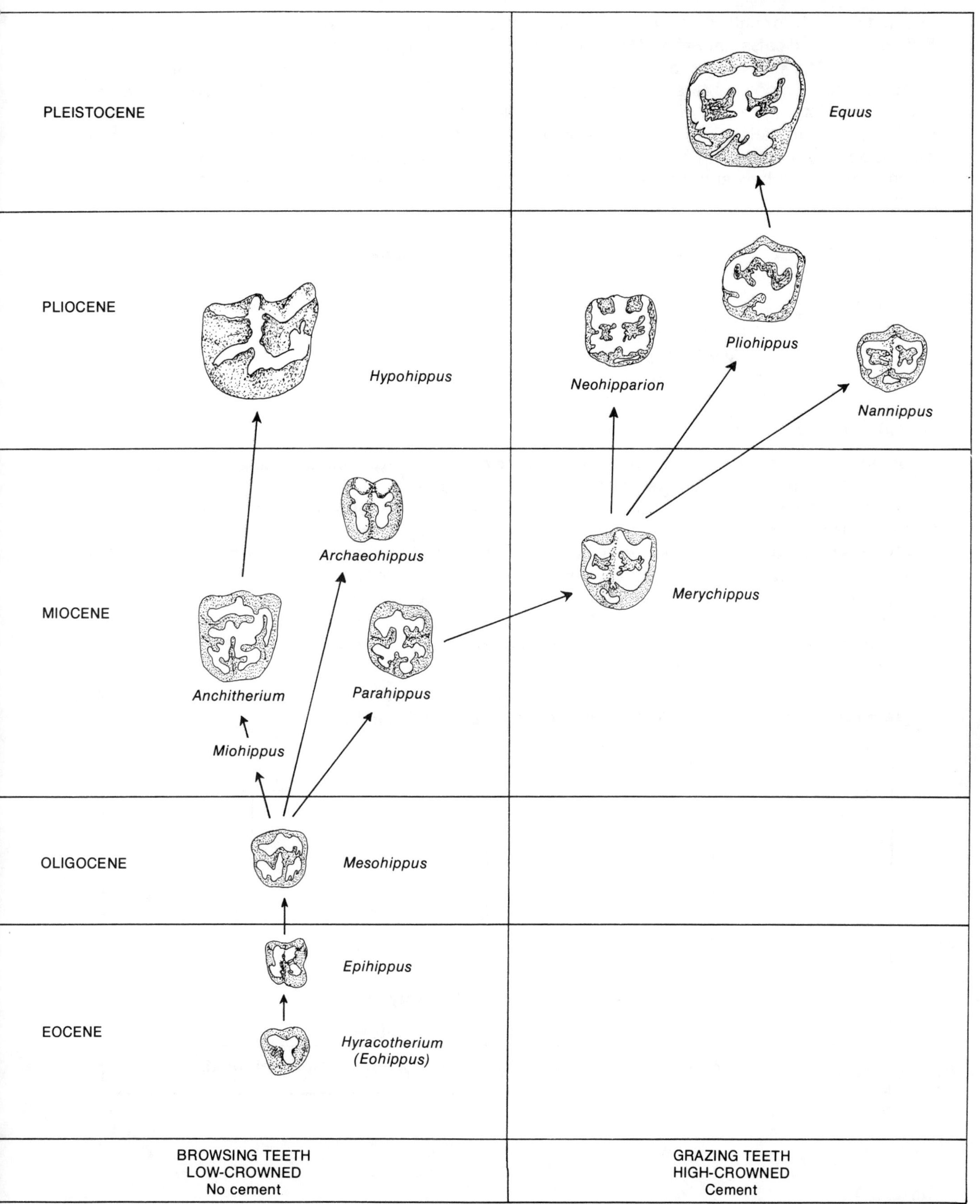

Figure 16.4. Stages in the evolution of horse teeth.

5. Additional information can be obtained from a graph such as the one you made. A nearly horizontal line indicates little change in the characteristic being studied (a slower rate of evolutionary change). The greater the deviation from the horizontal, the greater the rate of evolutionary change. Branching points represent genera where there would be a great deal of variation in population characteristics — a sort of "experimental" time. If, for example, one measured individual teeth from different populations of various genera, the percentage of variation could be calculated. Such calculations are recorded for four different genera. Referring to this data and to your graph, do you see any correlation between the percentage of variation and the evolutionary trends shown by the graph?

Genus	**Percentage of Variation**
Hyracotherium	17%
Merychippus	23%
Neohipparion	16%
Hypohippus	36%

Lower Limb Structure

Compare the leg bones of the modern horse to the models of legs from several fossil horses. Refer to the following brief descriptions as you look at the models.

In *Hyracotherium,* most of the body weight was carried on the four-toed, padded (dog-like) feet. The fourth toe was reduced in the hindlimbs. In *Mesohippus* only three padded toes carried the body weight. In *Merychippus* three toes were present, but most of the weight was carried on the larger central hoof. Bones of the toe were connected by strong, elastic ligaments which gave the foot a spring-like action as the animal ran. More recent horses possess only one functional toe, with a springing action. A one-toed, spring-footed limb is capable of rapid movement over hard surfaces. Does this gradual change in limb structure suggest anything to you relative to the environment in which these various horses lived?

Diagram these forelimbs and indicate approximate size beside each.

Hyracotherium 4-toed	*Mesohippus* 3-toed	*Merychippus* 3-toed	*Equus* 1-toed
Pad-footed		Spring-footed	

NATURAL SELECTION

Geologic evidence indicates that a series of environmental changes occurred during the Miocene in the area of North America and Europe where the horse family evolved. The large forested areas of the Eocene and Oligocene were gradually replaced by grassy prairie. This trend toward formation of flat, open grassland continued through the Pliocene. Grass, in addition to being normally covered with dust and grit, contains silica, a hard, sand-like substance.

Based on your knowledge of *Hyracotherium,* its environment, and subsequent environmental changes, correlate at least two major evolutionary trends in the Equidae with the selective factors operating.

Major Evolutionary Trend	Probable Selection Factor(s)
____________________	____________________
____________________	____________________

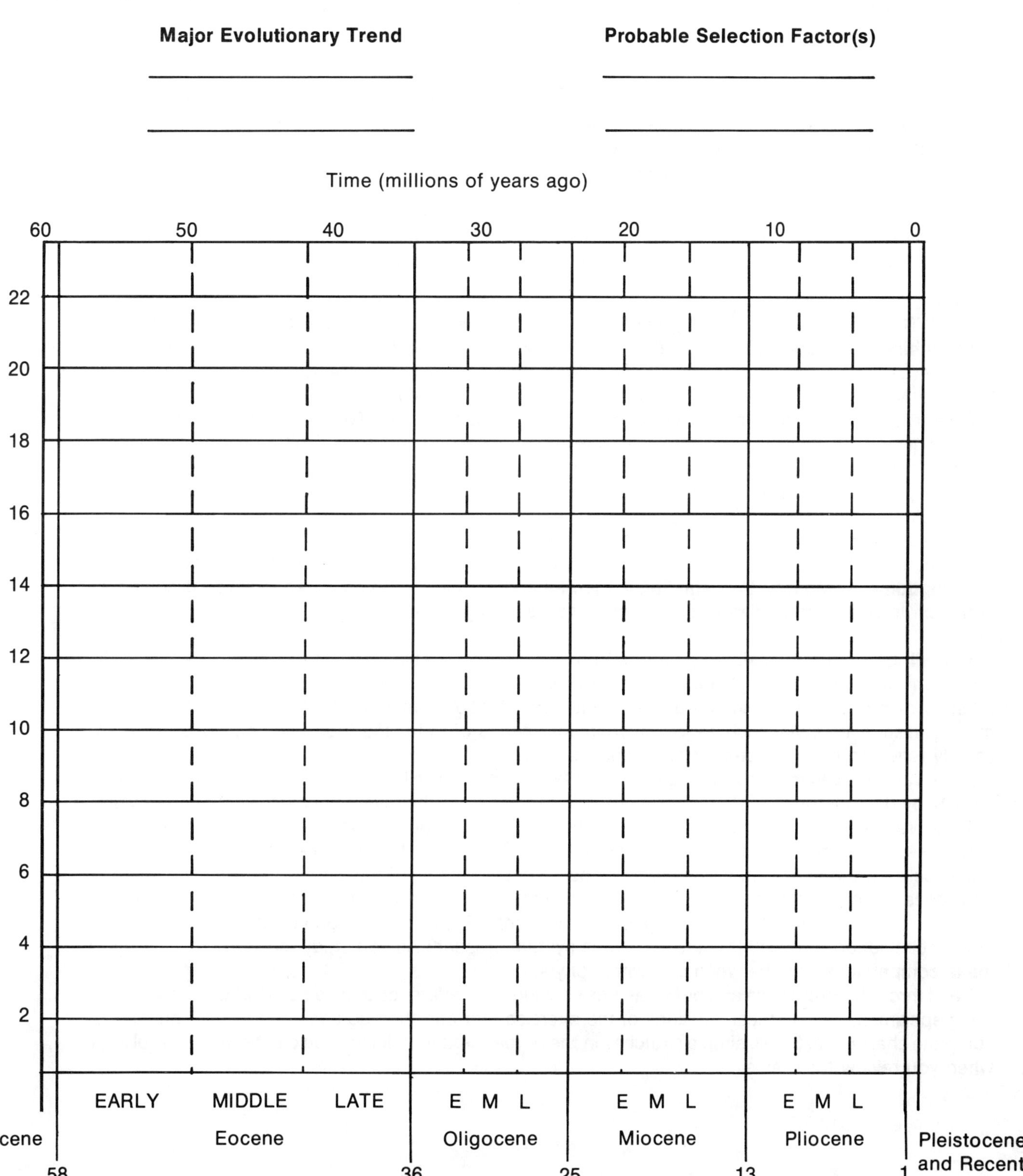

THE CLASSIFICATION OF LIVING THINGS

KINGDOM
  PHYLUM (DIVISION)
    CLASS
      ORDER
        FAMILY
          GENUS
            SPECIES

Classification is an attempt to group organisms which are closely related in their evolutionary origin. To undertake this grouping one utilizes information gained from studies of comparative anatomy, comparative biochemistry, paleotology, development, etc.

The unit of classification for all organisms is the species. A definition of this term applicable to all life is difficult, but it is usually defined as follows:

> **Species** — a population of individuals with similar structures and functional characteristics, which in nature breed only with each other and which have a common ancestry.

Closely related species are grouped together in the next higher unit of classification, the **genus** (plural, genera). The scientific name of an organism consists of two words, the genus and the species. The two-name or **binomial system** of nomenclature was first used consistently by Linnaeus. As an example, the domestic cat's name *Felis domestica* or Felis domestica applies to all varieties of tame cats (Persian, Siamese, Manx, etc.). Related species of the same genus would be *Felis leo* (the lion), *Felis tigris* (tiger), and *Felis pardus* (leopard). The dog belongs to a different genus, *Canis familiaris.* Note that the generic name is given first and is capitalized. The species name is second and is not capitalized. The two words must also be italicized or underlined. A genus name may be used alone if referring to all of the species making up that genus. The species name is not used by itself. It is really a modifying adjective and is meaningless when used alone.

Related genera are grouped into a single **family,** related families into a single order, etc. Different classification schemes have been proposed by various workers, but we will use the scheme involving five **kingdoms** — Monera, Protista, Fungi, Plantae and Animalia. The basic reasons for assigning an organism to one of these five kingdoms are discussed in your text.

You will be held responsible for general phylum characteristics. Attempt to associate a specific organism with each phylum. By thinking of this organism you can usually remember some of its characteristics, many or most of which will be representative of that phylum. Also, it helps to know something of the organism's life style, such as its environment, means of obtaining food, etc., because these are usually reflected in its structural characteristics.

In the various exercises several organisms will be studied in detail, while some phyla will be represented only by demonstrations. Demonstrations will also illustrate the diversity which exists in a phylum.

To appreciate other organisms it is necessary to understand the significance of their special adaptations. Keep this in mind as you study each organism. Is it multicelled or single celled? If it is multicellular, are its cells arranged to form organs or systems? What is the role of this particular type of symmetry? How does the organism obtain its food (specialized structures or behavior)? Does the digestive tract have one or two openings? How does gas exchange occur? How are nitrogen wastes excreted? What type of circulation is found? Is there a nervous system present? As you study the various phyla try to make comparisons both between and within phyla.

Read through the exercise and put an asterisk beside each reference to a demonstration, slide, model, living specimen, etc. Do these sections of the exercise first and then go back and answer questions, complete charts, etc. Small, simple sketches in the border next to a discussion can be very helpful later when you review for exams.

LABORATORY 17
MONERA, PROTISTA, AND LOWER PLANTS

Kingdom Monera
- Division Schizomycetes — bacteria
- Division Cyanobacteria — blue-green bacteria

Kingdom Protista
- Division Chrysophyta — golden algae, diatoms
- Phylum Mastigophora
- Phylum Sarcodina
- Phylum Sporozoa
- Phylum Ciliata

Kingdom Plantae
- Division Chlorophyta — green algae
- Division Rhodophyta — red algae
- Division Phaeophyta — brown algae

KINGDOM MONERA

The Monerans include two groups of organisms: the Division Schizomycetes (bacteria) and the Division Cyanobacteria (blue-green bacteria). Monera are differentiated from the rest of the living organisms by the procaryotic organization of the cells. **Procaryotes** have no nuclear membrane, no membrane-bound organelles, and nuclear division is by an irregular split rather than a true mitosis.

Division Schizomycetes (Bacteria)

Bacteria are Monerans which lack chlorophyll a. They are of tremendous importance to man, particularly in their role as decomposers. By this action organic matter (litter) is converted to simpler forms which can be used by other organisms. In this way elements necessary to life are recycled. Bacteria also function in **nitrogen fixation,** the process of converting gaseous nitrogen to nitrates and ammonia, forms usable by plants. Bacteria are also beneficial in food processing and in chemical production. The negative aspects of bacteria include their role in disease, e.g., pneumonia, diphtheria, syphilis, gonorrhea and tuberculosis. Bacteria as decomposers can also be harmful when what they decompose is of value, as food, clothing, etc.

1. Bacterial cultures are on display. Note odor, pigment (color), pigment diffusion, etc., and record this information.

Organism	Colony Color	Colony Shape	Other Unique Features

2. **Antibiotic Sensitivity Test** (demonstration) — Two bacterial genera have been tested for their susceptibility to a number of antibiotics. Paper discs impregnated with an antibiotic are placed on a nutrient medium, the surface of which has been inoculated with bacteria. Failure of the bacteria to grow around the disc indicates that organisms are killed or inhibited by the antibiotic. Such inhibition may or may not occur in the patient's body *(in vivo).* If the organism grows to the edge of the disc, the bacteria are resistant to that antibiotic and it would not be useful in treatment of an infection caused by that organism. Sketch results seen.

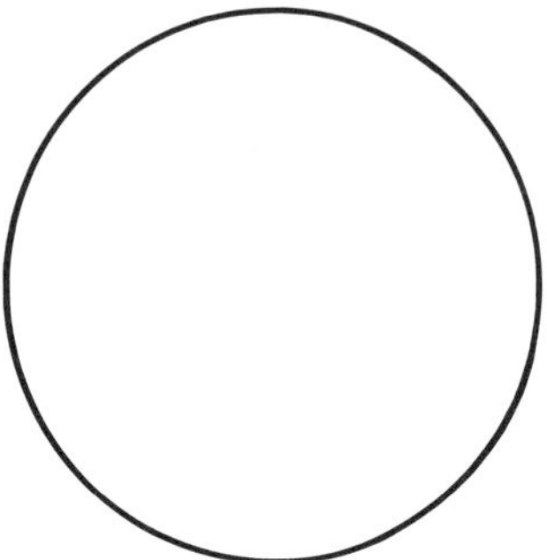

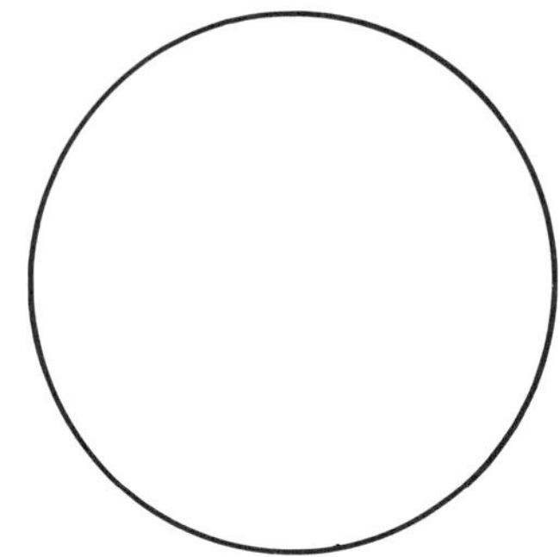

3. Observe the prepared slide and sketch the three major types of bacterial morphology: **coccus** (sperical), **bacillus** (rod shaped), and **spirillum** (coiled).

Coccus

Bacillus

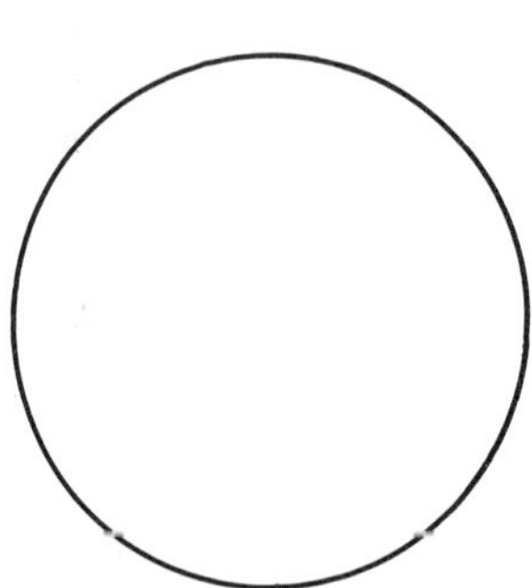

Spirillum

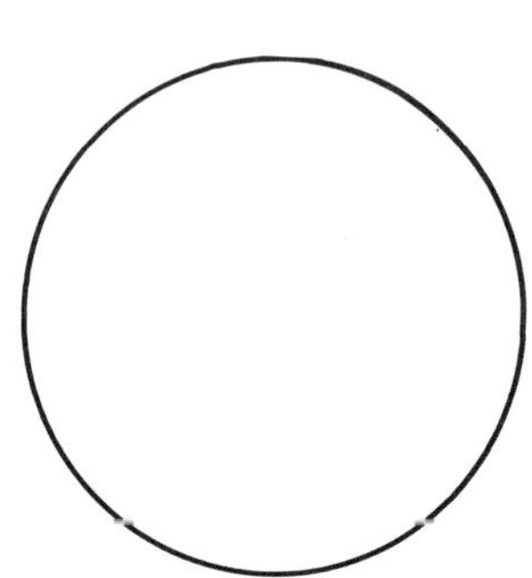

Division Cyanobacteria (Blue-Green Bacteria)

Procaryotic organisms with chlorophyll a, the blue-green bacteria, are important as **producers** — forming the base of many **food chains.** This group is abundant in freshwater habitats and also commonly occurs as a green covering on moist soils. In both habitats, they are important as nitrogen fixers. Often blue-green bacteria function as pioneer organisms on new soil or soil which has been exhausted of its nutrients by excessive farming. On the negative side, blue-green bacteria sometimes become numerous in reservoirs, causing a foul taste to the water or, in rare cases, producing a toxin.

1. Prepare a slide from the available specimens and sketch.

2. Are procaryotic organisms ever multicellular?____________________

3. How can one distinguish between blue-green bacteria and other bacteria?____________

____

4. Blue-greens often act as pioneer organisms. Why might they be more successful at this than other organisms?____________________________

5. In your text read about and be able to define primary producer, food chain and nitrogen fixation.

KINGDOM PROTISTA

Members of the Kingdom Protista are single-celled eucaryotes. Some are free-living heterotrophs, others are parasitic heterotrophs, and some are autotrophs. Plant-like protistans include the photosynthetic eucaryotes — dinoflagellates, golden algae and diatoms. Animal-like protistans are sometimes known as protozoans or "first animals." These include amoeba, foraminifera, paramecia, sporozoans such as malarial parasites and the flagellated organisms which cause sleeping sickness. Euglena are organisms with both plant and animal-like characteristics.

Division Chrysophyta (Diatoms)

Diatoms are characterized by their golden-brown color. Diatoms are said to live in "glass houses." These silica impregnated walls are often very intricately ornamented with pores, ridges and grooves. Diatoms, an extremely important component of the plankton, probably conduct most of the photosynthesis (and therefore oxygen production) in the earth's waters.

Observe the diatom specimens, noting particularly the symmetrical walls. Sketch a few.

Phylum Mastigophora

This phylum includes what are generally thought to be very primitive protistans. Many are important parasites, and all may be identified by their dominant feature — flagella. Name a disease caused by these organisms.____

These flagellates inhabit the blood of vertebrates and are carried from one host to another by blood-sucking invertebrates (usually insects).

Phylum Sarcodina

The sarcodinians are believed to have been derived from flagellated organisms through adaptations that involved loss of flagella and modification to allow the formation of projections or **pseudopodia** (false feet).

1. Observe a prepared slide of *Amoeba proteus* and sketch its structure. Note the **pseudopodia, cell membrane, ecto-** and **endoplasm, nucleus** and **vacuoles.** Pictures in your text can help you locate these if you have problems.

2. Two functions of pseudopodia are____________________ and __________

____

3. On what do free-living amoeba feed? ____________________

4. Some amoeba are parasitic and many of you may have first-hand understanding of how harmful some of them can be. *Entamoeba histolytica* produces a severe diarrhea (occasionally death) in man. The condition has been called Teheran tummy, Baghdad belly, Montezuma's revenge, etc.

5. Some sarcodinians secrete elaborate shells. See demonstration slides and text for examples of **foraminiferans** (limy shells) and **radiolarians** (silica shells). Of what importance are these protistans geologically?____________________________________

If they have shells, do they also have pseudopodia?__________

Phylum Sporozoa

This group lacks locomotor organelles (except for male gametes). They are parasitic and frequently show complex life cycles. Name a medically important (especially in tropical areas) sporozoan caused disease.____________________________________

Phylum Ciliata

1. The unicellular organisms in this phylum are remarkably complex, possessing numerous organelles specialized for particular functions. As an example we will use paramecia. Place a drop of the paramecium culture on a slide. Add a drop of methyl cellulose (to slow movements) and then a coverslip. Note the motion — forward, rotating, swerving. Is there a special manner in which this organism avoids obstacles?____________________________________

____

2. Note the motion of the cilia on the outer surface **(pellicle).** What are two functions of cilia?

____

3. Referring to your text, locate the following structures in **Figure 17.1** as you locate each on the specimen: **oral groove, food vacuole, contractile vacuole, nuclei, cilia, pellicle, cytopharynx.**

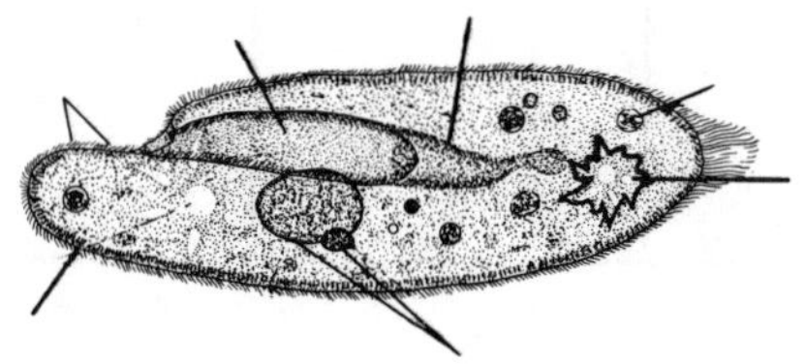

Figure 17.1. Paramecium, Phylum Ciliata, Kingdom Protista.

4. What is the function of the contractile vacuole?

5. What structures correspond to the digestive system of higher animals?

6. Does digestion occur intra- or extracellularly?

7. Some of these structures may be seen more readily on the prepared, stained slides available in the lab.

8. Review important features of this Kingdom later by completing the summary sheet.

THE PROTISTA — A SUMMARY

Characteristics of the Kingdom:					
Comparative study of vital processes	Division Chrysophyta	Phylum Mastigophora	Phylum Sarcodina	Phylum Ciliata	Phylum Sporozoa
Locomotion by					
Nutrition: How ingested		Same as Ciliata			Same as Ciliata
Where digested		Same as Ciliata			Same as Ciliata
How solid wastes eliminated		Same as Ciliata			Same as Ciliata
Metabolism: Respiration		Same as Ciliata			Same as Ciliata
Excretion of nitrogenous wastes		Same as Ciliata			Same as Ciliata
Examples					

KINGDOM PLANTAE

Organisms in the Plant Kingdom have **eucaryotic** cells with well defined nuclei, a nuclear membrane and chromosomes containing both DNA and histones. When the nucleus divides, it goes through a true mitosis. The cytoplasm contains a variety of membrane-bound organelles including chloroplasts, mitochondria, Golgi apparatus, etc.

In this laboratory session you will be looking at the lower plants, the algae, fungi, bryophytes and lichens. All lack vascular tissue and are usually smaller than the vascular plants which will be discussed next week.

THE ALGAE

The term **algae** is an artificial taxonomic term for organisms representing about ten divisions. Algae are mainly aquatic, lack vascular tissue, have simple reproductive structures and are **autotrophic.** Only four algae divisions will be studied. Some specimens will be living, some are mounted herbarium sheets and some are on prepared microscopic slides.

Division Chlorophyta (Green Algae)

This is the group from which the higher plants probably evolved. These algae share with the higher plants chlorophylls a and b, pigments which give both groups the "grass green" color. Green algae are abundant in freshwater habitats and are common in marine habitats. Within the group are found a great diversity of forms, including unicellular forms, filaments and sheets of cells.

Observe the specimens provided and sketch each. Be certain to note the prominent chloroplasts in microscopic forms.

Division Rhodophyta (Red Algae)

The red algae, almost exclusively a marine group, are most common in tropical waters. Often the red algae are large, easily visible to the naked eye. The color, from bright pink to red-brown, is due to the accessory pigment, **phycoerythrin.**

Observe the herbarium sheets. Note the larger size and the color, and sketch some examples.

Division Phaeophyta (Brown Algae)

The brown algae, commonly called kelps, are mainly a marine group but are most common in colder waters. These are the largest of the algae and some of the largest organisms in existence, attaining a length of 35 meters. The plant body is often differentiated into distinct regions, the **holdfast,** the **stipe** and the **blade.**

Observe and sketch some of the herbarium specimens, noting the differentiated regions.

QUESTIONS

1. Most algae are members of the plant kingdom and plant usually means non-motile. Does this characteristic apply to all the algae observed?

____

2. Red and brown algae have large plant bodies but lack vascular systems. How is water supplied to individual cells? _______________________

How is food supplied to individual cells? _______________________

3. How do the algae in the plant kingdom differ from the blue-green algae? _______________

____

4. Define autotrophic. _______________________________

____

LABORATORY 18

BRYOPHYTES (NON-VASCULAR) AND TRACHEOPHYTES (VASCULAR PLANTS)

Kingdom Plantae
- Division Bryophyta — bryophytes
  - Class Hepaticae — liverworts
  - Class Musci — mosses

- Division Tracheophyta
  - Subdivision Psilopsida
  - Subdivision Lycopsida
  - Subdivision Sphenopsida
  - Subdivision Pteropsida
  - Subdivision Spermopsida
    - Class Cycadae
    - Class Coniferae
    - Class Angiospermae
      - Subclass Monocotyledoneae
      - Subclass Dicotyledoneae

DIVISION BRYOPHYTA (BRYOPHYTES)

Bryophytes are non-vascular, autotrophic plants which are mainly terrestrial. Although living on land, their distribution is limited by their ability to supply all cells with water. Based on this description, in what type of habitat would you most likely find them?________________________

Would you ever find one the size of a VW?________________ Why? ______________________

____

Two classes within the Bryophyta are the **Hepaticae** (liverworts) and the **Musci** (mosses). Liverworts typically grow close to the ground and are **thallose** rather than leafy. Mosses, on the other hand, usually have an upright portion and are composed of "stems" and "leaves." Technically, these structures are not true stems and leaves because they lack vascular tissue. Bryophytes are more complex than algae in that reproductive organs are multicellular and the embryo remains within the female structure as it develops. The green plant which we recognize as a moss or liverwort represents the **gametophyte** generation. Each cell has a nucleus which is haploid *(n)*.

Male and female reproductive structures develop on the gametophyte plant. When water is available, sperm swim to the egg and fertilization results in a zygote *(2n)*. While still in the female structure (the **archegonium),** the zygote divides mitotically to form the embryo which grows into the **sporophyte** *(2n)* generation. The mature sporophyte generation is a stalked capsule standing above the leafy gametophyte. In the capsule, meiosis occurs and the resulting haploid spores become dispersed and grow into a gametophyte generation.

Compare **Figure 18.1** (the life cycle) with **Figure 18.2** (stages in the cycle), label each structure as *n* or *2n* and indicate where meiosis and fertilization occur.

Observe demonstration material of these two classes.

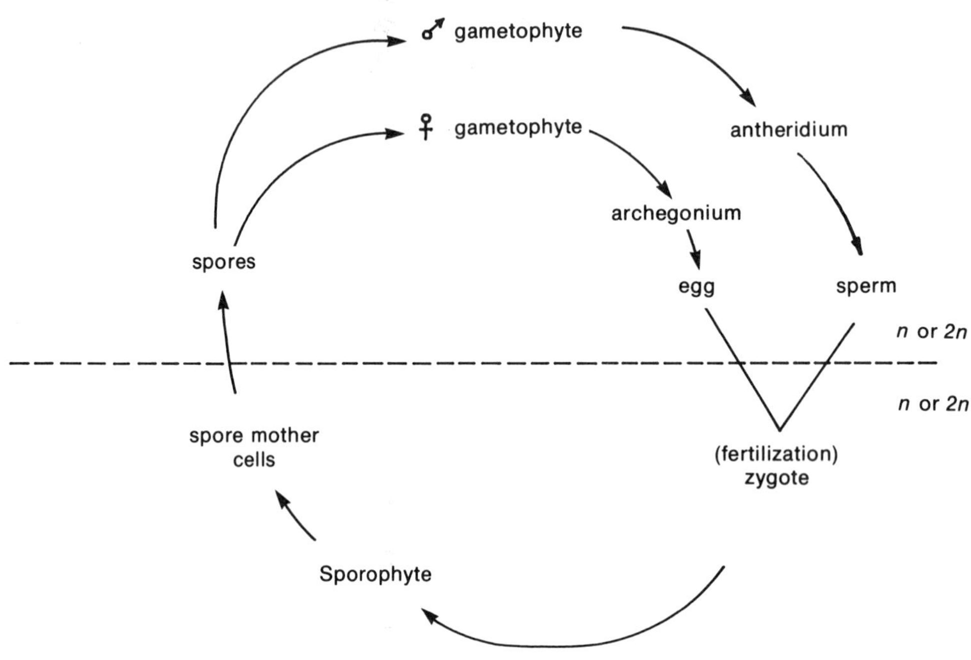

Figure 18.1. Life cycle of a moss.

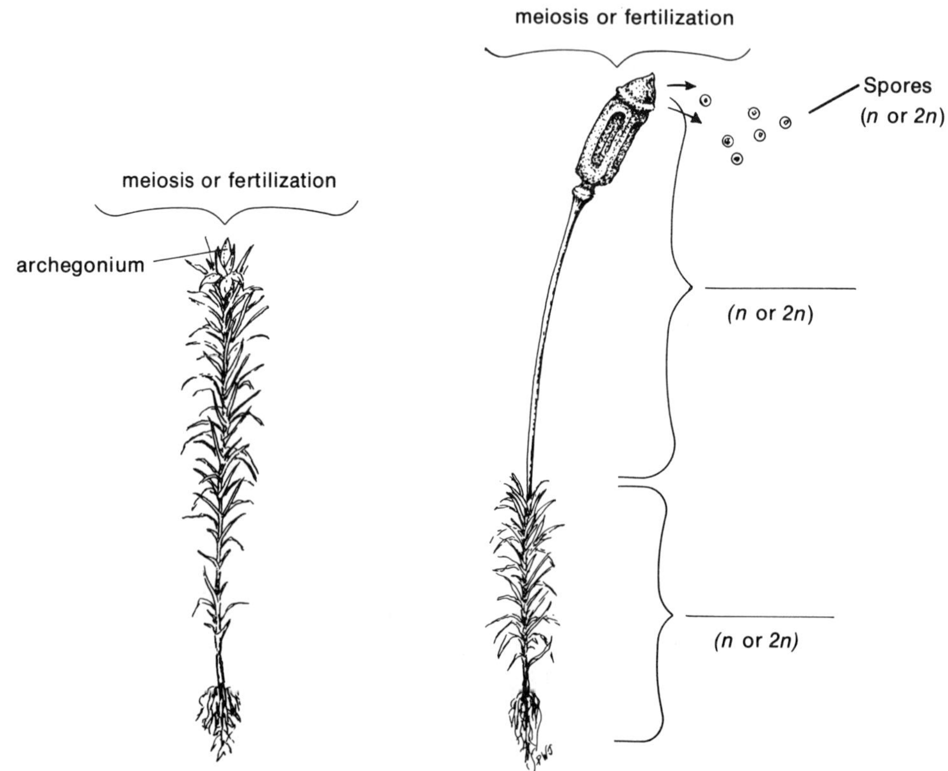

Figure 18.2. Moss stages.

QUESTIONS

1. Why do the bryophytes seem to be limited in size by a lack of vascular tissue while some algae attain much larger sizes? ______________________________

____

____

COMPARISON OF GROUPS

Indicate a "+" if that characteristic is found, a "−" if it is lacking, and an "S" if it occurs sometimes.

	cell walls	**vascular tissue**	**motile**	**chlorophyll "a"**	**procaryotic (P) or eucaryotic (E)**	**terrestrial (T) or aquatic (A)**
Schizomycetes	S		S			
Cyanobacteria	+		S			
Chlorophyta	+		S			
Rhodophyta	+		−			
Phaeophyta	+		−			
Bryophyta	+					

Is there a trend in motility? ______________________________

In aquatic vs. terrestrial habitat? ______________________________

THE VASCULAR PLANTS

The vascular plants are the dominant terrestrial plant group. Mosses, liverworts, lichens and fungi are numerous on land, but usually are of diminutive size. In aquatic habitats algal divisions are more successful and represent the dominant plant groups.

Vascular plants have achieved a great degree of freedom from the aquatic environment, a habitat from which they are believed to have evolved. Well developed vascular tissue provides a system for long distance transport of water, food and inorganic nutrients. Organisms lacking this system, e.g., the mosses, are restricted in size and never attain a height of more than a few inches. The surface of the vascular plants is covered by a waxy cuticle. What might be the function of such a cuticle?__________

____

LOWER VASCULAR PLANTS

The Division Tracheophyta may be split into five subdivisions: **Psilopsida, Lycopsida, Sphenopsida, Pteropsida** and **Spermopsida.** The first three are remnants of groups formerly important in the world flora. Much of the vast oil and coal deposits of the Carboniferous period (about 300 million years ago) are the remains of partially decomposed members of these three subdivisions. Vascular plants of today are mainly Pteropsida and Spermopsida.

Observe the demonstration plants (and sketch any unfamiliar ones) as you read the following characteristics of these vascular plant groups. Be able to recognize structures included in the descriptions. Some examples are shown in **Figure 18.5.**

Psilopsida

Only two genera in this subdivision are found in the modern world flora. These are thought to be closely related to the first vascular plants. The unique features of this group are the complete lack of leaves and roots. The plant body is a branched upright stem and an underground stem. **Sporangia** (spore producing structures) are found at the tips of the aerial branches.

Lycopsida

Lycopsida is represented today by several genera including the club mosses and the resurrection plant. The stem is usually covered by many small leaves, each having a single vein. Reproduction is by **spores** borne in the **axils** of the leaves. These are sometimes clustered at the tip of the stem.

Sphenopsida

Only one genus, *Equisetum,* remains of this large ancient group. However, this genus is widespread and well adapted to the modern world. The stem is unusual in being jointed. A whorl of small, one veined leaves encircles each node. Spores are borne in a **strobilus** at the tip of the stem. The stem of *Equisetum* is ridged and has a high concentration of silica. This material gives it a coarse feel, hence the common name "scouring rushes." Rub your fingers on the stem of this plant.

Pteropsida

Ferns, a very successful group of land plants, generally have large, highly subdivided leaves with numerous veins. Before maturity, leaves are curled into a tight coil called a fiddlehead. Fern **spores** are on the underside of the leaves.

As you observe representatives of the four subdivisions of lower vascular plants, complete the following table.

	locations of spores	**arrangement of veins**	**major photosynthetic structure**
Psilopsida			
Lycopsida			
Sphenopsida			
Pteropsida			

The order of these groups (Psilopsida to Pteropsida) represents a progression from more primitive to more advanced forms. In this progression, what structure becomes highly developed?

____

SPERMOPSIDA — SEED BEARING PLANTS

The Spermopsida are the most successful group of the vascular plants. In this group, seeds are produced as the main reproductive units. A **seed** includes an immature new plant (embryo), stored food and a protective coat (Exercise 15). Seeds are much more complex reproductive units than the single celled spores. A wide variety of plants are found in this subdivision including the classes Cycadae, Coniferae and Angiospermae. Representatives of these classes are on display. Again, simple sketches of any unfamiliar plants may be helpful.

Cycadae — the Cycads

The cycads are not abundant, but are easily recognized by their large, coarse compound leaves. These usually attach densely to a short woody stem. Seeds are produced on cones.

Coniferae — the Conifers

The conifers comprise a large, widely distributed group, unique in that all members are woody. Leaves are relatively small and either needle-like or scale-like. Seeds are produced on cones. The woody cones which we recognize as "pine cones" are the female cones. Male cones are small, produced in the spring, and last only a few weeks. During this period, male cones produce tremendous quantities of pollen which are released as yellow clouds.

1. By what characteristics can cycads and conifers be distinguished? ______________________

____

2. Cycad leaves are most like those in what other group? ______________________

Angiospermae — the Flowering Plants

Today, the dominant terrestrial plant group is the Angiosperma. Flowering plants have been more successful than the other plant groups in such diverse habitats as desert, tropical forest, tundra, grasslands and savannah. In a few habitats, such as subarctic forests and western mountainous regions, the Coniferae often dominate.

Angiosperms are identified most easily by two important features — **flowers** and **fruits.** Both features are involved in reproduction and may be partially responsible for the success of this class.

The flower is a shortened branch of highly modified leaves, specialized to promote pollination. In many angiosperms, the large, colorful flower attracts animals (especially insects) which spread the pollen. In others, flowers are modified to facilitate pollination by the wind.

The basic **flower** is composed of four **whorls** or layers of flower parts. The outermost whorl is made up of **sepals.** Often the sepals are green and function mainly to protect the flower before it opens. In other cases, they are colorful. The **petals,** the next whorl, are usually the most colorful parts and function mainly in attracting pollinators. Inside the petals are the **stamens,** the male reproductive structures. The stamen consists of a supporting stalk and pollen producing sacs called **anthers.** The innermost part of the flower is made up of the female **pistils.** Each pistil contains a swollen base **(ovary),** a neck **(style)** and a landing surface for the pollen **(stigma).** After fertilization, the pistil develops as the **fruit.** Inside the ovary are one or more **ovules** which develop into seeds.

1. Label flower parts in **Figure 18.3.**

2. Observe the flowers on demonstration. Record the name of the flower and determine the number of units in each whorl.

Whorl	Flower 1	Flower 2	Flower 3
sepals	________	________	________
petals	________	________	________
stamens	________	________	________
pistils	________	________	________

3. Angiosperms can be subdivided into two subclasses, the Dicotyledonae (dicots) and the Monocotyledonae (monocots). **Monocots** are characterized by (1) flower parts in threes or multiples of three, (2) narrow leaves with veins running parallel to each other, and (3) one cotyledon in the seed. **Dicots** have (1) flower parts in multiples of fours or fives, (2) veins in a net pattern, and (3) two cotyledons in the seeds. (See **Figure 18.4.**) Of the flowers just observed, which would be monocots?

____

______________________ Dicots? ______________________

4. The possible evolutionary relationships of the plants studied in the last two exercises are given in **Figure 18.5.**

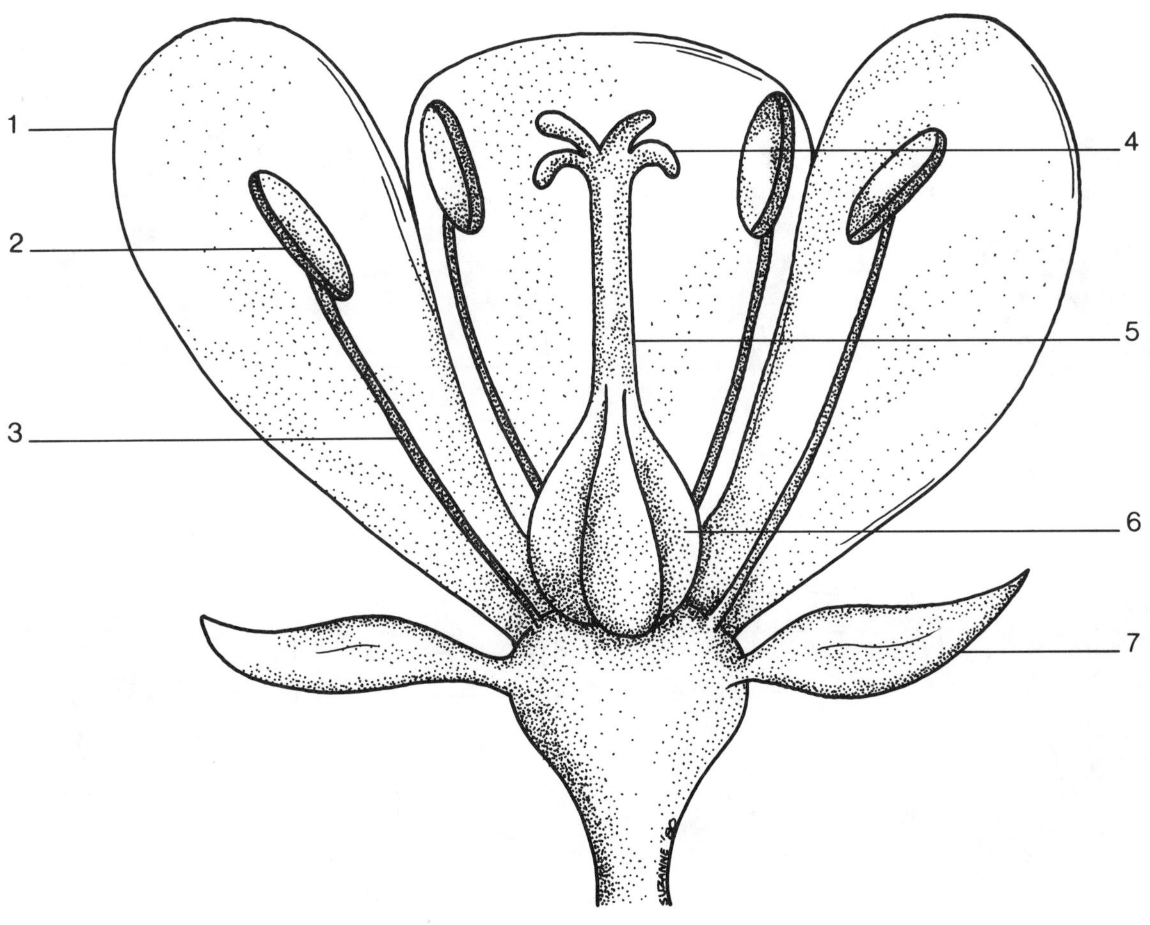

#2 + #3 = ____________________

#4 + #5 = ____________________

Figure 18.3. Typical flower parts.

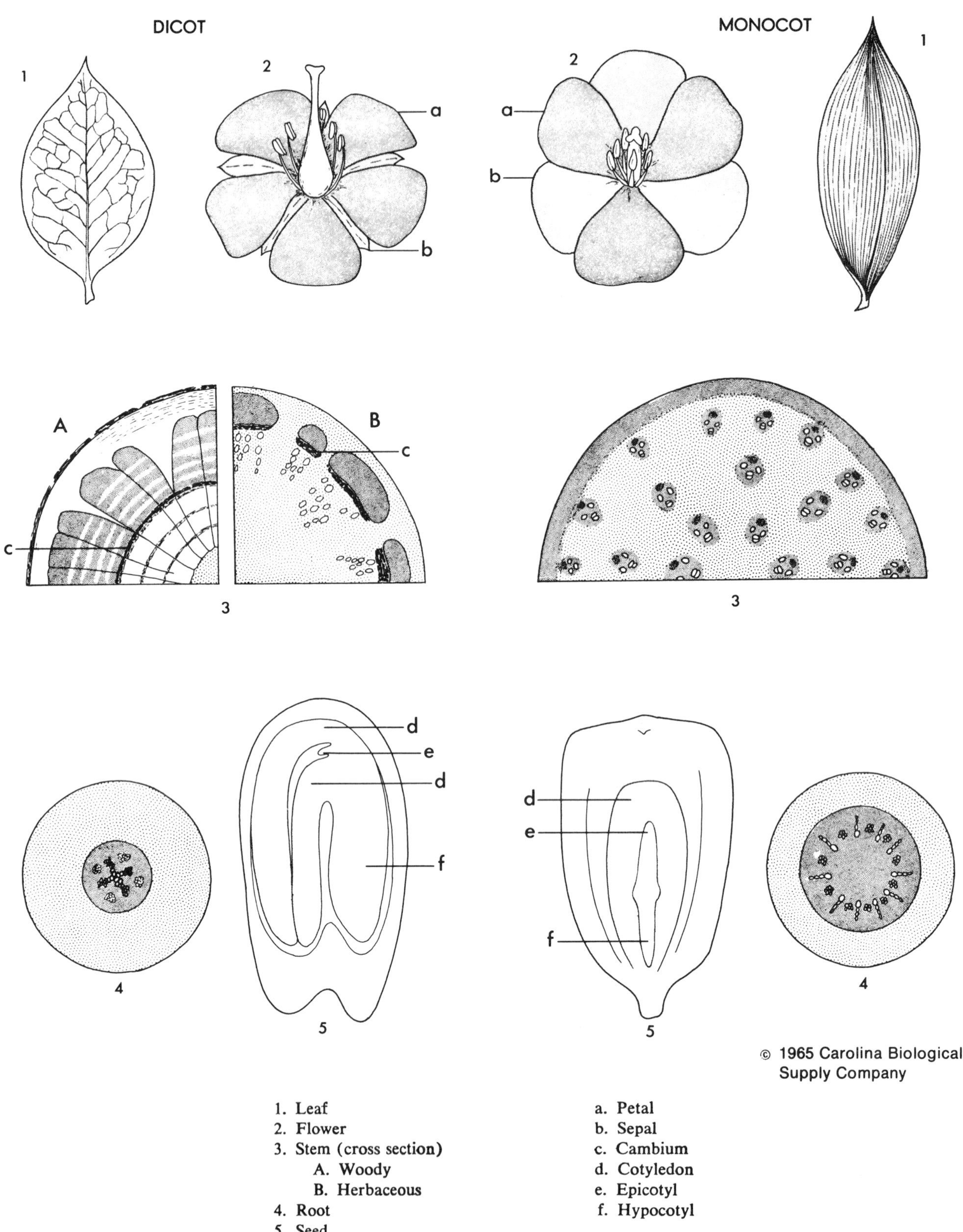

1. Leaf
2. Flower
3. Stem (cross section)
   A. Woody
   B. Herbaceous
4. Root
5. Seed

a. Petal
b. Sepal
c. Cambium
d. Cotyledon
e. Epicotyl
f. Hypocotyl

Figure 18.4. Comparison of monocot and dicot characteristics.

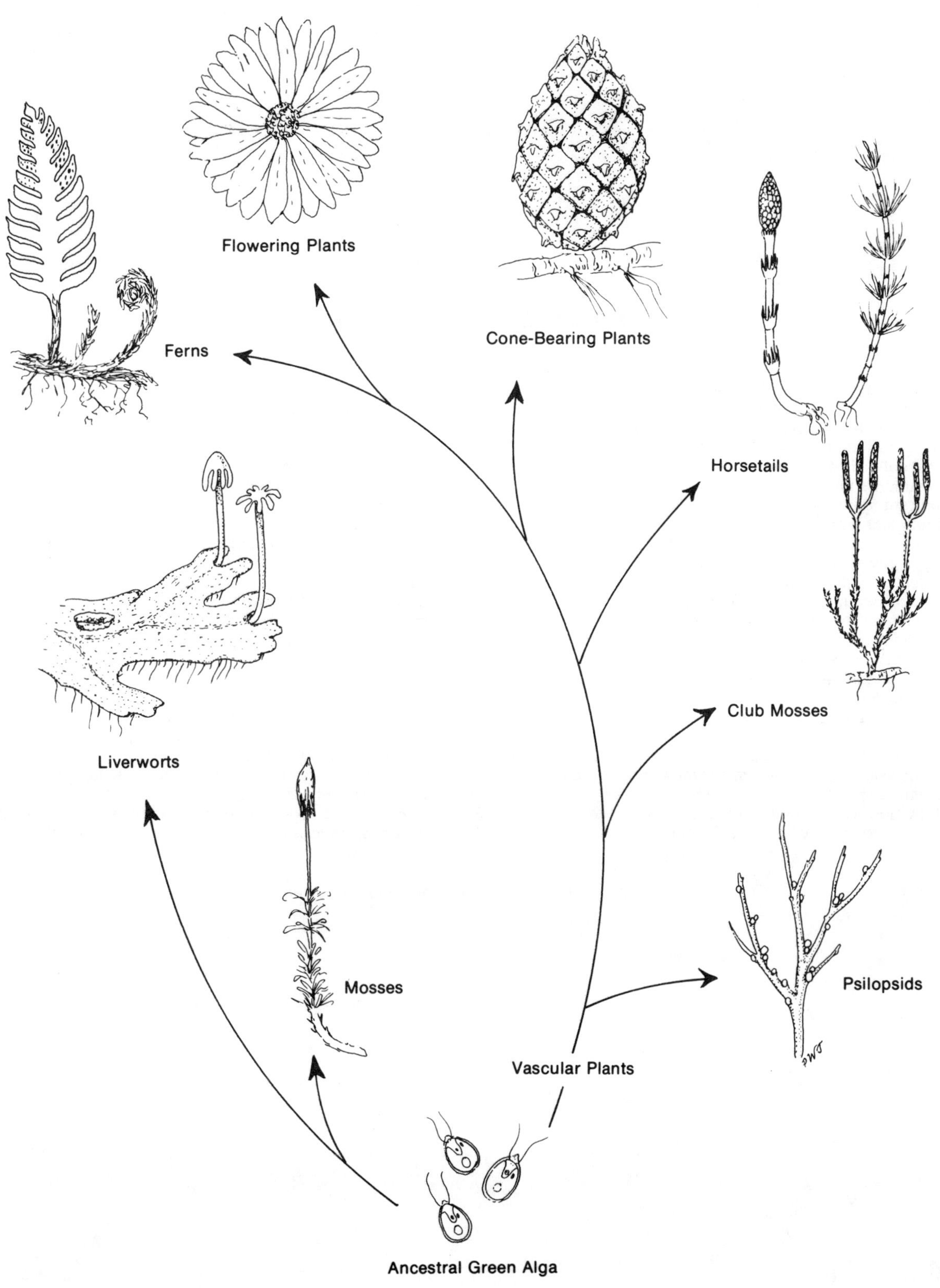

Figure 18.5. Possible evolutionary relationships of green plants.

VASCULAR PLANTS CROSSWORD PUZZLE

Across

1. dominant terrestrial plant division today
8. modified leaf on a conifer
9. Xylem supplies structural___to trees. (backward)
11. landing surface for pollen (in a flower)
14. oldest subdivision studied
17. fern subdivision
19. important feature of angiosperms (backward)
20. smooth leaf margin
21. leaf margin with wave-like arrangement
23. have highly subdivided leaves with spores on underside
24. swollen base of the pistil
25. ___buds occur where petiole joins stem
27. leaf with one petiole and one blade
30. A rosebush is a di___
32. the immature new plant in a seed
34. once abundant — have large, coarse, compound leaves and reproduce by seeds in cones
35. The leaves are___te if the petioles come off like the letter "Y."
36. scouring rush or *Equi*___
37. Use the axillary___to determine whether it is a simple or a compound leaf.
38. subdivision with jointed stems:___sida
39. main reproductive unit of the Spermopsida

Down

1. vascular tissue function
2. found in cycads and firs, but not in oaks (singular)
3. Anthers produce___en.
4. the flowering plants
5. colored part of a flower (singular)
6. Xylem and phloem comprise the___tissue.
7. Corn is a___cot.
10. ovary + style + stigma =___l
12. Net veins and flower multiples of 5 would belong to a___cot.
13. pines, firs, spruces
15. outermost whorl of a flower (singular)
16. stamen = stalk +___r
18. thin section joining leaf to stem
22. Leaflets are subdivisions of a leaf___.
23. Basic___are composed of 4 whorls.
25. Leaves may be opposite or___.
26. Seeds contain___in the cotyledons.
27. the subdivision of seed bearing plants:___sida
28. Toothed and lobed refer to leaf___gins.
29. resurrection plants and club mosses:___sida
30. Many leaflets indicates a___leaf.
31. The embryo and food are surrounded by a seed___ (backward.)
33. High concentrations of silica in the stem are characteristic of the scouring___. (singular)

(see page 209 for correct answers.)

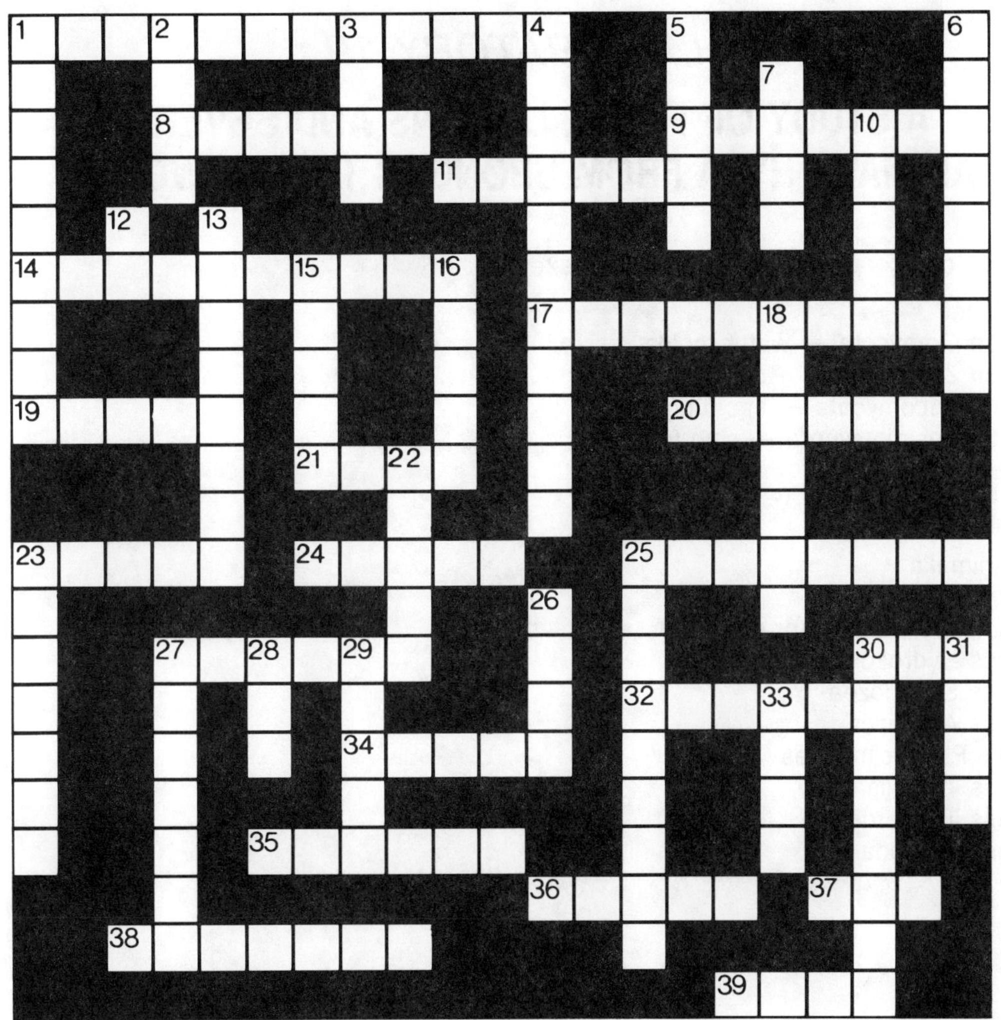

VASCULAR PLANTS CROSSWORD PUZZLE

LABORATORY 19

A STUDY OF FUNGI, LICHENS AND SEVERAL ANIMAL PHYLA FROM SPONGES TO FLATWORMS

Classification categories to be studied in this exercise:

Kingdom Fungi
- Division Oomycota — water molds
- Division Zygomycota — conjugation fungi
- Division Ascomycota — sac fungi
- Division Basidiomycota — club fungi

"Lichens"

Kingdom Animalia
- Phylum Porifera
- Phylum Coelenterata (or Cnidaria)
  - Class Hydrozoa
  - Class Scyphozoa
  - Class Anthozoa
- Phylum Platyhelminthes
  - Class Turbellaria
  - Class Trematoda
  - Class Cestoda

KINGDOM FUNGI

Fungi are organisms lacking chlorophyll and vascular tissue. Fungi function in the **ecosystem** primarily as decomposers, acting with the bacteria to break down organic litter. When the food source is a living organism, they are called **parasites,** e.g., ringworm, athlete's foot and Dutch elm disease. When the food source is non-living, they are called **saprophytes.** Fungi are also important in food processing (especially yeasts in the brewing and baking industries), as a direct food source (mushrooms, truffles and morels), and as disease organisms (especially in plants).

A short slide presentation will be given to illustrate the various fungi described below. In addition, plastic blocks and preserved and living specimens are on display.

True Fungi

These fungi are differentiated from slime molds by the presence of cell walls. True fungi are either unicellular (e.g., yeasts) or composed of filaments of cells called **hyphae.** Fungi reproduce by spores, which are produced by the millions. A common component of the biotic portion of the air is fungal spores. On the proper food source, fungal spores quickly germinate and start digestion of the food. As an example, recall how quickly bread mold appears on bread even though fungi inhibitors have been added. The four main divisions, the Oomycota, the Zygomycota, the Ascomycota and the Basidiomycota, are divided primarily on the basis of differences in methods of sexual reproduction and spore formation. Members of each class are on display. Sketches might prove helpful later.

In the **Oomycota** the plant body is composed of filaments which lack cross walls. As a result, hyphae are long, multi-nucleate tubes. Common Oomycotas include water mold (a common disease organism of fish) and other disease fungi such as potato blight and downy mildew of grape. A common **Zygomycota** is bread mold.

The **Ascomycota** are called the sac fungi because the sexually produced spores are formed in sacs called **asci.** Most of the group are large fungi such as morels and cup fungi, but also included in the group are unicellular yeasts.

In the **Basidiomycota** (club fungi) sexually produced spores are formed on club shaped cells called **basidia.** The more familiar fungi belong to this group: the mushrooms and toadstools, bracken fungi and coral fungi. Included in this group are many serious plant pests including wheat rust and corn smut. In both the Ascomycota and the Basidiomycota the hyphae have definite cross walls.

QUESTIONS

1. What differentiates the fungi from the algae? ____________________

____

from the bacteria? ____________________

THE LICHENS

Lichens are difficult to classify because each plant is a **symbiotic** association between an alga and a fungus. The two components grow and reproduce as a unit. The resulting plant (the lichen) has properties which neither component by itself has. In the symbiotic relationship, the algal component uses photosynthesis to provide food. The fungal component forms the shape of the organism and provides protection while reducing water loss. Lichens seem well adapted as pioneer organisms, often colonizing rocks, barren soil and wood. As they grow, organic matter and soil accumulate to eventually form a habitat in which other plants can survive. The pale blue-green crust seen on local tree trunks and branches is usually a lichen.

Observe the lichens on display.

KINGDOM ANIMALIA

The animal kingdom contains multicellular heterotrophs capable of motion and exhibiting some degree of cellular specialization. The kingdom is sometimes divided into two major groups: **parazoa** — animals lacking organized tissues and a digestive cavity (sponges) and **metazoa** — those with organized tissues and a digestive cavity. Metazoans may then be further subdivided into (1) animals with radial symmetry (jellyfish and hydra), (2) protostomia — animals in which the mouth develops from the blastopore (flatworms, roundworms, annelids, molluscs and arthropods) and (3) deuterostomia — animals in which the anus develops from the blastopore (echinoderms and chordates). The advent of multicellularity provided advantages, but also presents problems. What are some of the advantages? Some of the problems that had to be solved?

Specific organization of parts in an individual animal is known as **symmetry.** Organisms may be arranged like cylinders with parts radiating similar to the spokes of a wheel **(radial symmetry).** Many can be divided into two halves, each a mirror image of the other half **(bilateral symmetry).** Still others exhibit no symmetry **(asymmetrical).** You will see examples of these symmetrical types in the various animals to be studied in the next few weeks.

Below is a list of terms you will need. These terms are illustrated in **Figure 19.1.** Others will be added later.

anterior This region usually contains the head; it is directed forward in locomotion.
posterior The tail end or end opposite the head.
dorsal The back or upper side.
ventral The belly or underside, opposite the dorsal side.
lateral The side.
medial The midline, halfway between the right and left sides.
proximal The near end, that is, the part nearest the body; for example, the elbow is proximal to the hand.
distal The far end; the toes are distal to the ankle.
peripheral The outer edge.
oral The mouth side.
aboral Opposite to the mouth side.
longitudinal section A plane parallel to the body's long axis.
cross (transverse) section A plane at right angles to the longitudinal section.

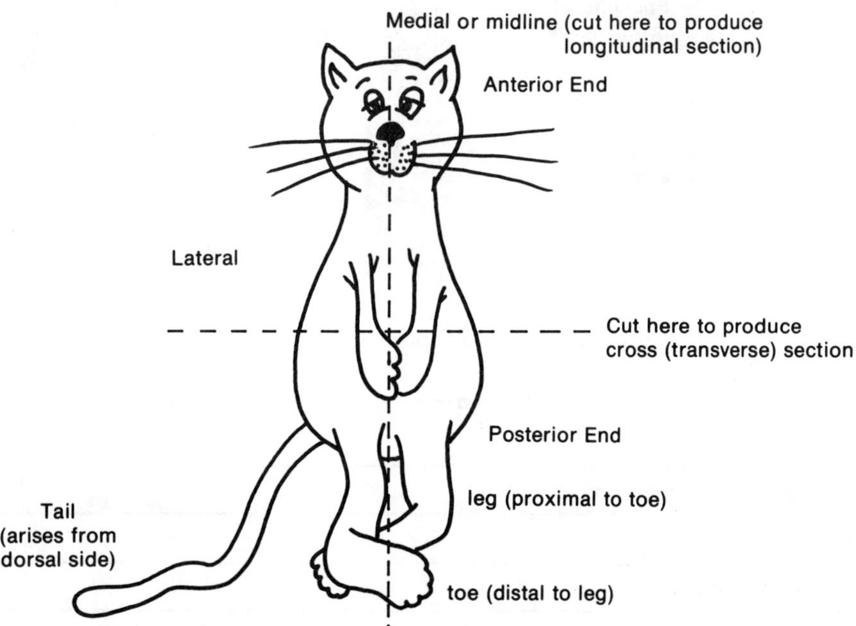

Figure 19.1. Ventral view of a bilaterally symmetrical cat.

PHYLUM PORIFERA (SPONGES)

Sponges are **diploblastic** (two-layered) animals. The layers, however, are not organized into true tissues, and sponges lack organ systems. The phylum is considered to be at the cellular grade of organization. As sponges show distinct cell types, division of labor among these cells, and specific structures, they are definitely above the protozoan colony, although certainly representing the most primitive existing multi-cellular animals. Most sponges are marine and are found at depths varying from a few inches to 3.5 miles. Form is maintained by "skeletons" or **spicules** of calcium carbonate (Class Calcarea) or silica (Class Hexactinellida), or by a network of tough protein fibers (Class Demospongiae).

Circulatory, digestive and respiratory systems are inseparable. Water from the environment is circulated throughout the body by a **pore** and **canal** system. In **Figure 19.2** note the **collar cells** (flagellated cells often used to suggest a possible evolutionary relationship between sponges and protozoans). Sponges engulf microorganisms (much as do the amoebas) and intracellular digestion occurs. Diffusion then distributes nutrients to neighboring cells. Aerobic respiration occurs by utilization of oxygen dissolved in the surrounding water. Using the last three statements as a basis —

Can sponges use large particles of food? __________ Can they grow as large as your lab instructor?

__________ Why? ____

Simple diffusion of metabolic wastes into the water solves the excretory problem. Reproduction may occur sexually or asexually (by budding). Sexually the sponge is **hermaphroditic** (capable of producing both sperm and egg). Fertilization results in the formation of free-living ciliated larvae which swim away, settle down and form mature sponges. Essentially no organized nervous or muscular tissue system is found in the sponges.

Examine demonstration specimens of various sponges, looking for **spicules, opening, point** of **attachment** and **pores.**

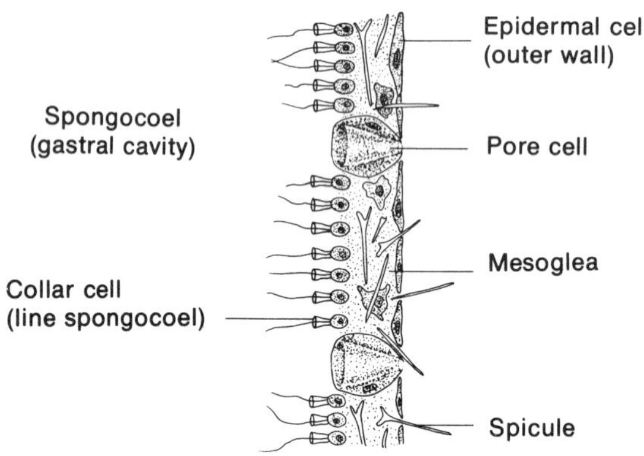

Figure 19.2. Section through the wall of a sponge (*Grantia).*

THE PORIFERA — A SUMMARY

Characteristics of the phylum:	
Symmetry	
Body wall construction	
Habitat	
Skeletal materials	
Locomotion	

PHYLUM COELENTERATA

The diverse animals comprising this phylum may occur in two forms — **polyp** and **medusae.** Polyps are tubular, with one end closed and the other end having a mouth surrounded by tentacles. Medusae are free-swimming, umbrella-shaped organisms. Two cell layers — an outer ectoderm and an inner endoderm — are present in all. In two of the three major groups we see the beginnings of a third layer called **mesoglea** (not a true tissue layer). The body plan is that of a sac-within-a-sac, forming a central digestive cavity. Symmetry is radial, and if a skeleton exists it is calcareous. **Nematocysts** (coiled poisonous threads used to paralyze prey) are a common feature.

Circulatory, digestive and respiratory systems are inseparable. Tentacles sweep unicellular organisms into the central cavity or trap organisms and bring them to the mouth. Digestion occurs extracellularly in this cavity into which enzymes are secreted. Respiration and waste excretion are solved by simple diffusion. True nerve cells are arranged in a nerve net, the forerunner of the more complex nervous systems which will emerge in more advanced phyla.

The sac-like body plan of the coelenterates resembles the early embryonic gastrula stage of higher animals with a single external opening (the embryonic blastopore, the coelenterate mouth) and a large internal cavity (embryonic archenteron, coelenterate gastrovascular cavity). Reproduction is complex — asexual only (budding), sexual only, or both.

Coelenterates are separated into classes according to the dominant embryonic state. During development, the animal progresses from a **hydroid (sessile** or attached) to a **medusae** (free-living) stage.

Class Hydrozoa (predominant polyp stage)

Hydra — a solitary hydrozoan

1. Hydra, slender organisms 1.0 to 2.5 cm long, live in fresh water attached to submerged rocks, twigs, etc. Refer to **Figure 19.3** and note structures **(bud, tentacles, mouth)** and different cell types.
2. Observe a prepared slide of a cross section of a hydra and identify **ectoderm** (outer cell layer), **mesoglea, endoderm** (inner cell layer), **gastrovascular cavity, nematocyst.** In what cell layer would you find digestive cells?____________________ What is the general shape of the body in cross section? ____________________

What type of symmetry would be associated with this general shape?____________________

3. Plastic blocks with embedded specimens are on demonstration.
4. Cellular structure in hydra is typical of coelenterates, but life cycle and certain other features are not. Many coelenterate polyps live in colonies, each colony containing individuals specialized for several different functions.

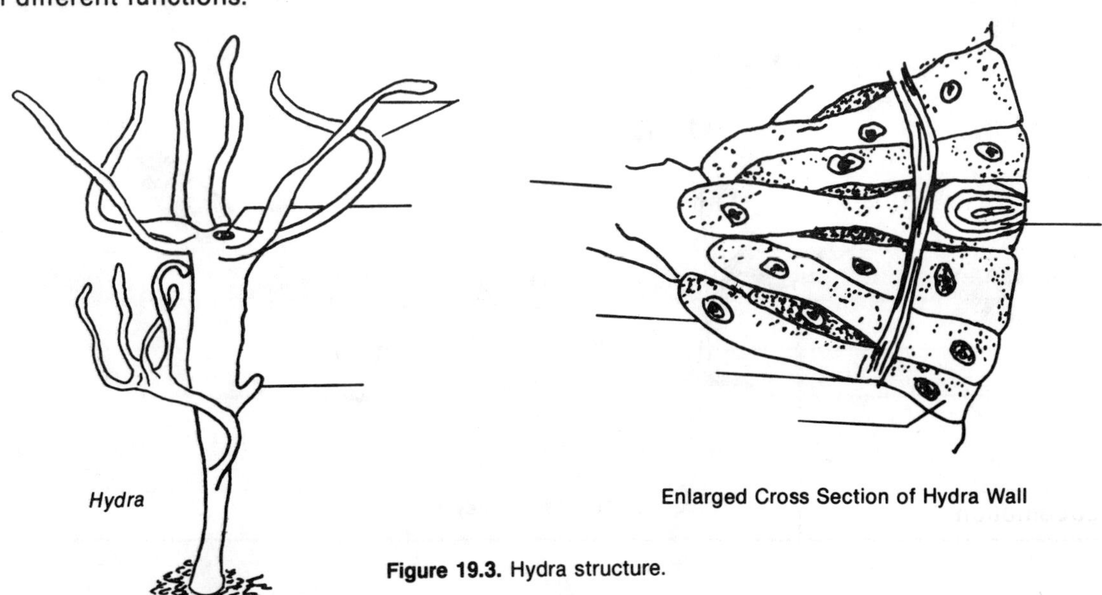

Hydra

Enlarged Cross Section of Hydra Wall

Figure 19.3. Hydra structure.

Obelia — a marine colonial hydrozoan

1. The life cycle of *Obelia* is diagrammed in **Figure 19.4.** This coelenterate includes a medusa in its life cycle. The medusa is umbrella-shaped, free-floating and usually produces gametes. Medusae are often formed by a polyp specialized for this purpose.

2. Obtain a prepared slide of an *Obelia* polyp stage. Locate and be able to label: **feeding** and **reproductive polyps, tentacles, medusae buds, perisarc** (hard, outer covering), **mouth, gastrovascular cavity.**

3. *Obelia* medusae are rather small to study in detail, so the large medusa stage of the next class will be observed instead.

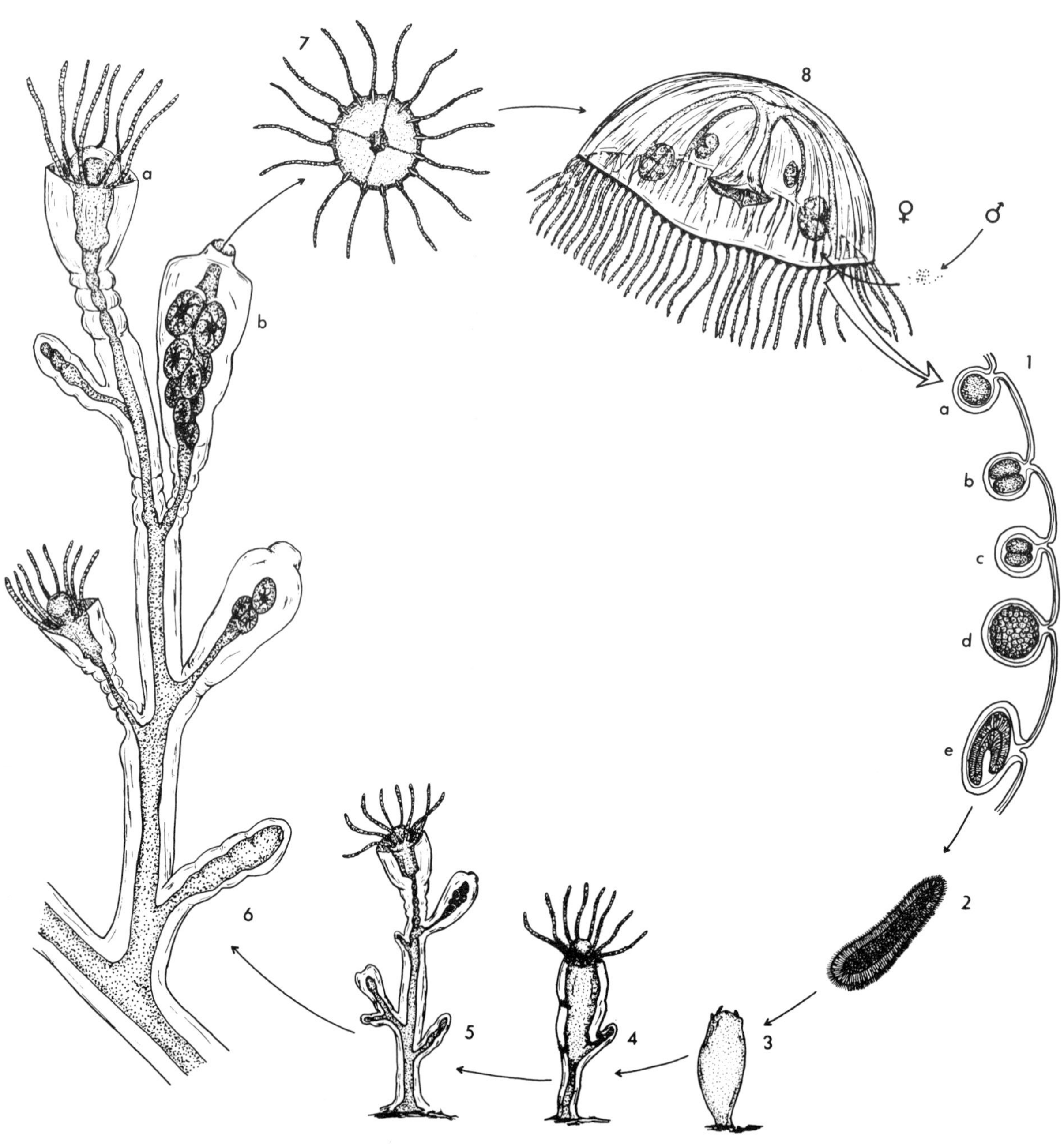

Figure 19.4. Hydrozoa life cycle.

Class Scyphozoa

In this class the free-living medusa predominates. *Aurelia* is a common jellyfish found almost throughout the world, frequently observed in large groups, drifting or moving slowly by rhythmic contractions of the shallow, saucer-shaped bell. Bell size varies from less than three inches to twelve inches.

Observe the specimen (preserved and plastic blocks) at each table and label **Figure 19.5** as you locate the various structures. What type of symmetry does it have? ______

What structures make up the fringe around the periphery? ______

On one surface find four long, tentacle-like structures **(oral arms)** attached to a fleshy process (manubrium) and surrounding the **mouth.** The mouth opens into a short gullet leading to a **stomach** (divided into four pouches). The brown, thickened edge of each pouch is a **gonad.** The whitish lines are **canals.** In each of the eight equally spaced indentations in the margin is a **sense organ.** These sense organs consist of an eyespot (sensitive to light), a hollow statocyst (for balance and direction in swimming) and two sense pits (probably sensitive to chemicals).

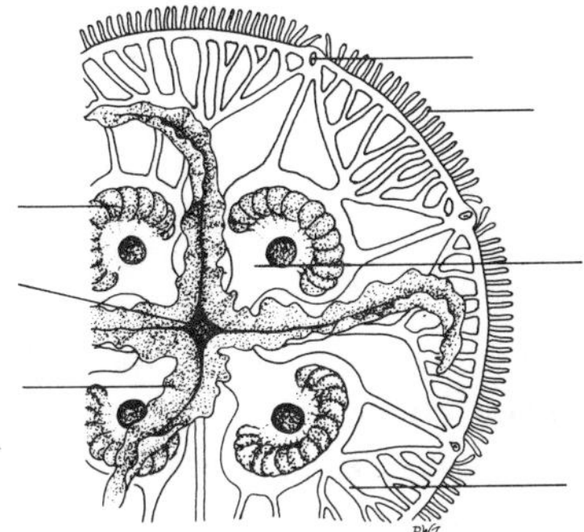

Figure 19.5. A portion of *Aurelia* medusa.

Class Anthozoa

Anthozoa, marine polyps which lack a medusa stage, have a gastrovascular cavity divided by a series of vertical partitions. These organisms are usually found in warm shallow waters, but a few occur in the polar regions and some are found at depths to 17,000 feet.

1. **Corals** are small, fragile polyps which are usually colonial. They secrete protective limestone skeletons and after a time huge reefs may build up. Examine demonstration corals and note the variety of form. Only the skeletons remain. Locate the small pockets which formerly contained the polyps.
2. **Sea anemones** are a highly specialized polyp type of coelenterate. They contain a well developed nerve net and several sets of specialized muscles. Observe the specimen of *Metridium.* Anemones occasionally attach to shells of crabs and molluscs and may thus be dispersed widely. Many living anemones are brilliantly colored.
3. Other coelenterates are on display.

PHYLUM PLATYHELMINTHES

You have already studied one member of this phylum (the liver fluke) as an example of extreme specialization for a parasitic life. In addition to its parasitic members, the phylum is also important because it illustrates several major advances (over previously studied phyla). Some of these **advances** are:

1. Establishment of a true mesoderm to form a **triploblastic** organism, and **organization of tissues** into definite organs and many organ systems.
2. **Bilateral symmetry** and **cephalization** (concentration of sensory organs at anterior end). Of what advantage is bilateral symmetry? ______
3. More **complex sensory organs** (eye spots) and **cerebral ganglia** ("brain").
4. **Muscle fibers** arranged in longitudinal, circular and oblique bundles. How is this advantageous? ______

5. Primitive excretory organs **(flame cells).**

6. **Specializations** for a parasitic mode of life: (a) degenerative digestive tract (Class Trematoda) or no digestive tract (Class Cestoda), (b) degenerative nervous system, (c) suckers and hooks for attachment to host, and (d) extensive, highly developed reproductive system.

Class Turbellaria (Free-Living Flatworms)

This group is illustrated by small fresh water planarians, commonly found in slow streams or ponds adhering to stones, twigs, etc.

1. On a prepared slide of a planarian observe the branched digestive tract leading from the pharynx. What is the function of the frequent lateral projections from the main branches?________

____

How many openings to the outside are found in this digestive system?__________Excretory and nervous system structures are difficult to observe due to the thickness of the animal.

Class Trematoda (Flukes)

Flukes are parasitic flatworms. Refer to Exercise 16 to relate structures observed at that time to the phylum characteristics.

Class Cestoda (Tapeworms)

Tapeworms, parasites with complex life cycles involving one or more intermediate hosts, are highly modified for reproduction, with both sexes found in the same worm **(monoecious** condition).

1. Obtain a slide of *Taenia pisiformis,* a common dog tapeworm. As you observe the following structures (with and without a scope) label **Figure 19.6.** The **scolex** is the knob-like head with a slender neck. The anterior end of the head forms the **rostellum** on which are a series of hooks. The wider portion of the head has a row of **suction discs.** What is the function of the hooks and suckers? ______________________________ Note the transverse segments beginning posterior to the discs. These segments are newly formed sexually reproducing **proglottids** (immature in this region). Was the fluke body segmented? ______________________

2. Use a low power scope to observe internal structures. With time these proglottids increase in size (mostly length) and develop sex organs. Each mature proglottid (second section on slide) contains numerous testes and a set of female organs. Self-fertilization does occur, but usually fertilization occurs between adjacent proglottids through a **genital pore** (located at the side of each segment). See your text for aid in labeling. After fertilization, male organs and many of the female organs disappear.

3. The third section on the slide is a **gravid proglottid** in which the fertilized eggs are fully developed. A mature proglottid may contain in excess of 100,000 eggs. Mature proglottids break off and are eliminated with the feces of the host. Eggs will hatch if eaten by an appropriate intermediate host. They encyst in the host tissues and complete development to the adult form only when eaten by the appropriate final host. The branching structure filling much of the interior is the egg-filled **uterus.** Nerve cords and excretory canals extend the length of the body on each side. Can you find an intestine?______Is the shape of the parasite in any way related to the lack of respiratory and circulatory systems?________Explain.

Summary

List five items in each of the following categories. Use **non-living** items commonly found in an American home or in this laboratory.

	Asymmetrical	Radially Symmetrical	Bilaterally Symmetrical
1.			
2.			
3.			
4.			
5.			

Important areas from this exercise would include: terminology, all questions, summary tables, structures, examples and characteristics of phyla and classes, advances seen in flatworms.

Two unidentified invertebrates are on display. Can you classify them?

Unknown #1 Phylum ____________________ Reason: ____________________

Unknown #2 Phylum ____________________ Reason: ____________________

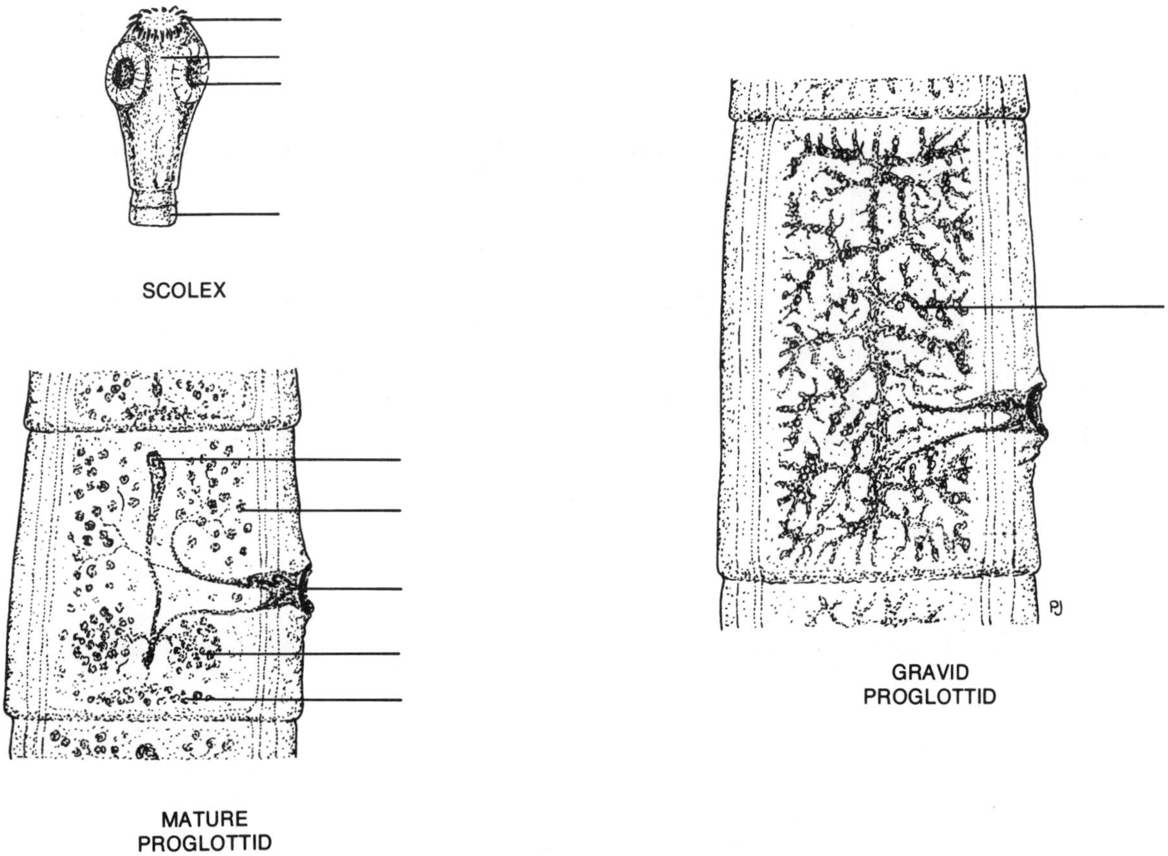

SCOLEX

MATURE PROGLOTTID

GRAVID PROGLOTTID

Figure 19.6. *Taenia pisiformis* structures.

COMPARATIVE STUDY OF COELENTERATA AND PLATYHELMINTHES

	Phylum Coelenterata			Phylum Platyhelminthes		
Item	Hydra	Jellyfish	Anemone	Planarian	Fluke	Tapeworm
Habitat						
Body Shape in Cross Section						
Number of Tissue Layers						
Type of Symmetry						
Type of Body Cavity	acoelomate	acoelomate	acoelomate	acoelomate	acoelomate	acoelomate
Digestive System						
Reproduction						

LABORATORY 20

SOME MORE ADVANCED INVERTEBRATES

It is strongly recommended that simple outline drawings be made of most of the specimens seen today in lab. Detailed works of art are unnecessary, but a simple diagrammatic representation is a help in remembering characteristics of the various phyla and classes. As you study the various animals ask yourself . . . "What do this animal's external features tell me about its way of life?"

Classification categories studied in this exercise are:

- Kingdom Animalia
  - Phylum Aschelminthes
  - Phylum Mollusca
    - Class Pelecypoda
    - Class Amphineura
    - Class Gastropoda
    - Class Cephalopoda
  - Phylum Annelida
    - Class Oligochaeta
    - Class Polychaeta
    - Class Hirudinea
  - Phylum Arthropoda
    - Class Chilopoda
    - Class Diplopoda
    - Class Arachnida
    - Class Crustacea
    - Class Insecta

PHYLUM ASCHELMINTHES

Major **advances** occurring in the Phylum Aschelminthes include: (1) a **complete digestive tract** (with two openings) and a **tube-within-a-tube pattern,** and (2) a **body cavity** (although not a true coelom).

Multicellular animals studied thus far have all been **acoelomates,** characterized by a solid cell mass instead of open spaces within the body. The Phylum Aschelminthes constitutes another line of evolutionary development — the **pseudocoelomates** (pseudo = false, coel = body cavity). For example, animals such as rotifers and nematodes contain spaces around the various organs, but these spaces are not lined with mesoderm. **Coelomate** animals have these spaces lined with mesoderm. Sexes are almost invariably separate **(dioecious)** in contrast to the monoecious platyhelminthes. Class Nematoda contains unsegmented roundworms with long, slender cylindrical bodies. Some are free-living in the soil and water and others are parasites in plants and animals. Nematodes lack circulatory and respiratory organs.

Ascaris lumbricoides

1. Adult *A. lumbricoides* worms inhabit the digestive tract of pigs, where semifluid intestinal contents provide food. Male and female worms copulate and the female releases fertilized eggs into the host's feces. When other pigs eat food contaminated with these eggs, eggs hatch in the pig intestine. Larvae burrow into the intestinal wall, enter capillaries and are carried to the lungs via the circulatory system. In the lungs they increase in size prior to migrating to the trachea and then the pharynx where they are swallowed. The cycle starts again in the intestine.

2. Sexes are separate, with the male shorter and more slender than the female. He also has a sharply curved posterior end and a pair of hair-like structures (used in copulation) extruding from the anal opening. Is your specimen male or female? ____________

3. Adult *Ascaris* have an outer non-cellular covering (cuticle). What is its function?

____

Trichinella spiralis

1. In man the condition known as trichinosis results from eating raw or partly cooked pork containing encysted larvae of *T. spiralis.* This parasite infects an estimated 17% of the U.S. population.

2. Examine the demonstration slide of a muscle section containing encysted *T. spiralis* larvae.

3. Name two other important nematode parasites of man.

____

PHYLUM MOLLUSCA (MOLLIS = SOFT)

Basically, molluscs are a marine group which have secondarily invaded fresh water and land. This is one of the largest and the most familiar of the invertebrate groups (80,000 living species and 35,000 extinct ones). Members of this group are found universally — underground, in mud flats, in coral reefs, in tree tops, in lakes, at snow line in the Himalayas, three miles deep in the ocean, as parasites, etc. Molluscs vary from a fraction of a gram to two tons in weight. Outwardly phylum members are quite different from each other, but all exhibit similar body plans and share some common features. Molluscs have two unique features — the ventral **muscular foot** (variously modified, but usually for locomotion) and a **radula** (rasp-like organ present in the pharynx of most). A special molluscan structure is the **mantle** — covering the soft visceral mass and usually secreting a hard, calcium-containing shell. All require a moist environment and most are aquatic. Land forms either stay in damp places (slugs) or conserve water by closing their shells when conditions are dry (snails). In land snails part of the mantle cavity is modified to act as a lung.

Both living and preserved specimens belonging to this phylum are available, as well as fossil representatives and shells. Other examples are shown in **Figure 20.1.**

Class Pelecypoda (Hatchet Foot)

In oysters and clams the basic molluscan plan has been modified into two shells **(bivalve)** held together with strong muscles. Gills are used for breathing and there is no head or radula. Feeding occurs by drawing water in through siphons and filtering out plankton on the cilia of the gills.

Examine and sketch various pelecypods on demonstration. Look for similarities and differences between members of this class and other molluscan classes.

Class Amphineura (Both Neurons)

1. Examine a specimen of the small marine mollusc, the chiton (also see **Figure 20.1**). Describe the general body shape and arrangement. Is it twisted or straightened? ______________

Flattened or rounded? ____

2. Observe the ventral surface. Can you identify the structure which occupies most of this surface?

____

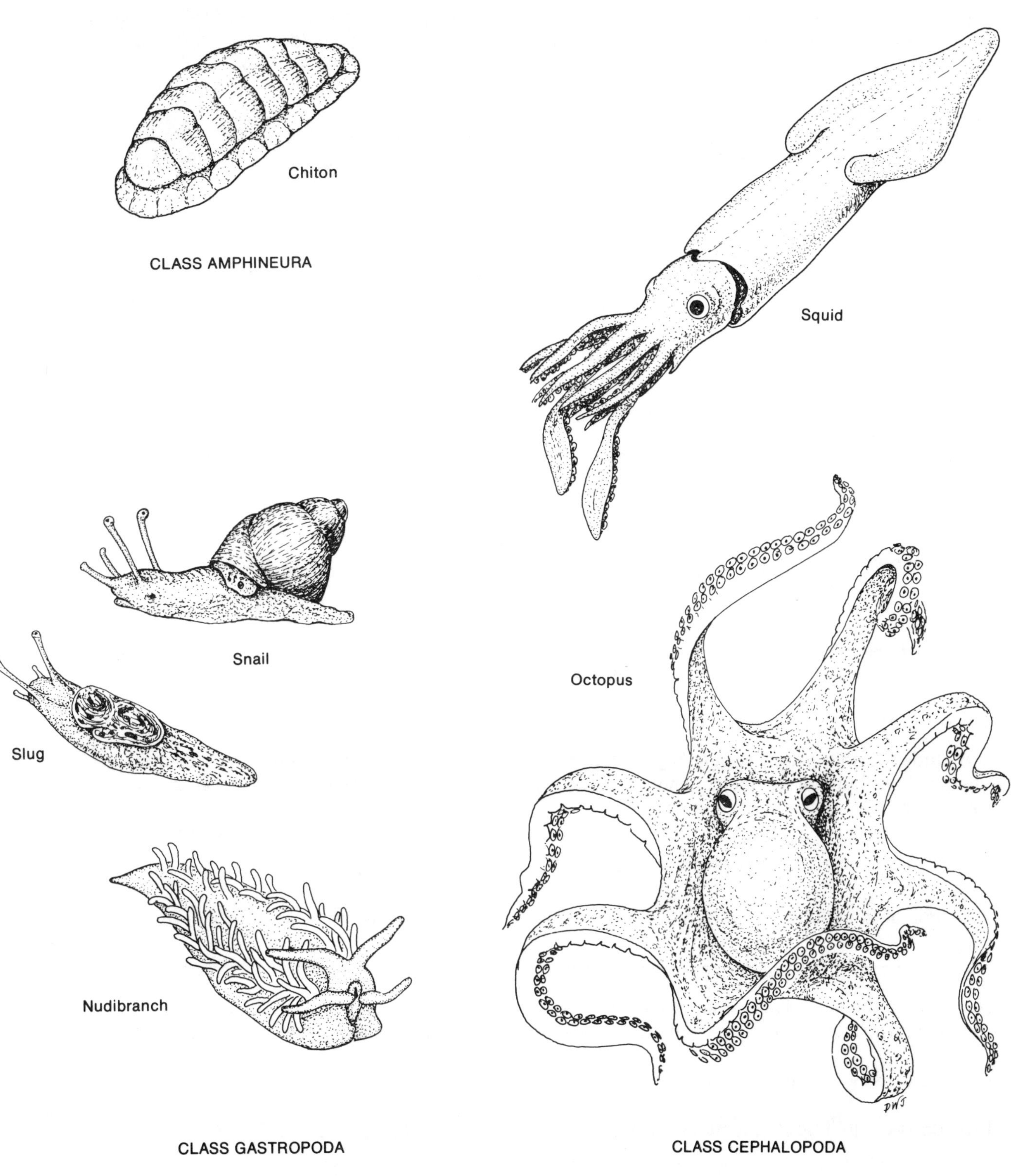

Figure 20.1. Representative classes of molluscs.

Class Gastropoda (Stomach Foot)

All terrestrial molluscs are in this class. The visceral mass is enclosed in a coiled shell during the early stages of development. In most gastropods the coiled shell is retained in the adult, but it has been lost in some species.

1. Living and preserved specimens are on demonstration. Note the various shell modifications. Also see **Figure 20.1.**

2. Observe a specimen of the slug, *Limax,* whose shell has been reduced to a small flattened plate covered by the mantle. On the basis of observation alone, would you be inclined to place this organism into this class or into some other class or phylum? ______________________________

Why? ______________________________

3. Observe the marine limpet *Acmea testudinalis.* How does this shell differ from that found in most other gastropods? ______________________________

4. Observe the freshwater snails in the aquarium. What means of locomotion do they use? Can you see a foot? ____________ a head? ____________ tentacles? ____________

Reach into the aquarium and detach a snail from the side of the glass.

What happens to the body? ______________________________

What happens to the opening? ______________________________

Snails may not be very impressive (except perhaps to another snail), but they are really rather advanced animals. The visceral mass contains a complete digestive system, a nervous system and a circulatory system including a heart enclosed in a true coelom.

Class Cephalopoda (Head Foot)

Included in this class are very highly developed marine molluscs with an internal or an external shell, a modified foot (tentacles) and a head with prominent, highly developed eyes. These molluscs are active, free-swimming organisms. Also see **Figure 20.1.**

1. Examine a squid. What morphological characteristics do you see in this animal that might be advantageous to a predatory type of existence? ______________________________

This animal is capable of rapid movement. Could this be related to its lack of an external shell?

Explain ______________________________

A vestige of the shell does remain as a horny plate buried in the visceral mass. The mantle covers all of the ventral side of the animal including the large mantle cavity where water is held for jet propulsion. How many tentacles are there? ____________ Are they all alike? ____________

How does a squid hold objects or food? ______________________________

The eye is quite complex — complete with cornea, iris fold, lens and posterior optic chamber. The mollusc eye originates embryonically from the skin in contrast with the composite brain-skin eye of the vertebrates, and since the vertebrates arose independently this is an excellent example of **convergent evolution.**

2. Examine an octopus. How many tentacles does it have? ________ In what ways is it like and in what ways does it differ from the squid?

3. Also included in this class are the **nautilus** and **ammonites.** Ammonites are extinct cephalopods whose fossil remains are quite common in local limestones. Many of the specimens were formed from mud casts or impressions of the shell's interior. The shell was coiled in a flat plane and the soft body of the animal was contained within it. As the animal grew, chambers were formed sealing off the posterior parts of the shell by means of **septae** or partitions. Many ammonites had very ornate septae (seen as complex lines on the fossils) and also ridges, knobs, etc., externally. Compare some of the patterns seen on these fossils.

Survey of the Mollusca

As you examine and sketch the various molluscs consider the following. Which lack a visible shell? Are scallops more closely related to snails or to clams? Which close relative of the squid still has an external shell? Which are edible?

PHYLUM ANNELIDA

Phylum members are called the segmented worms due to a structural arrangement based on a series of similar segments or **metameres.** Metamerism apparently was a prerequisite for the evolution of appendages and provides the organism with efficient, powerful, and independent muscle masses. Annelids possess a **tube-within-a-tube** body plan associated with a **true coelom.** Although lacking in the earthworm, some annelids possess paired appendages. Annelids living on the sea floor often have colorful gills for respiration. Phylum members possess **closed circulatory systems,** increased cephalization and **nephridia** which remove wastes from blood as well as from the coelom. **Metamerism** (division of the body into successive groups of somites or segments) makes its first appearance in the animal kingdom in this phylum. Have you studied somites in the lab before?

In what organism? ______________________________

Class Oligochaeta

1. The earthworm is really an atypical annelid, but it is familiar, easily obtained (and inexpensive!) — so — obtain an earthworm from the container and place it in a dish partly filled with water. Orient yourself as to the worm's **anterior** (more pointed) and **posterior,** and **dorsal** (darker) and **ventral** (lighter and flattened) sides. What type of symmetry is evident? ______________

2. Note the regular segmentation. The iridescent outer covering is the cuticle. With your fingers note the segmental bristles **or setae.** How many setae are there per segment? __________

3. The saddle-shaped structure is the **clitellum.** What is the number of the first segment located in the clitellum? __________

What is the clitellum's function? ______________________

4. In Exercise 7 you studied the cross section of the earthworm. Our primary interest at this time is to illustrate the appearance of a coelom and to know **why** it is a coelom. Refer to **Figure 7.3** to refresh your memory as to its structure.

Class Polychaeta

1. The sandworm, *Nereis virens,* is a typical marine annelid. It lives in burrows in the mud or sand at tide level, leaving this burrow at night to search for food.

2. On the specimen of *N. virens* note the segmental appendages or lateral outgrowths called **parapodia.** Are they all alike? ____________

What is their function? ____

3. Note the many bristle-like setae (characteristic of the group).

Class Hirudinea

Leeches live in water or moist places. Some attack and eat smaller animals and others are adapted to sucking the blood of vertebrates such as fish, turtles and mammals. A leech attaches with suckers, pierces the skin and gorges itself on blood, which can be stored for months (because the leech secretes chemicals which keep the blood fresh and clot-free).

1. Obtain a specimen of a leech. Make an outline drawing of a lateral and dorsal view.

2. Compare its morphology with that of *Lumbricus* and *Nereis.*

How does it differ? ____

How is it similar? ____

3. Is the body of the leech rounded or flattened dorsoventrally? ____________________

Is there a well defined head? ____________

4. The anterior sucker surrounds the mouth. The leech has three teeth. Can you suggest a function of these teeth? ______________________________

Also locate the posterior sucker.

5. The many transverse divisions are not considered true segments since they are superficial, and in fact exceed the number of true segments found internally.

PHYLUM ARTHROPODA (JOINTED FOOT)

Members of this group form a very diverse phylum with segments highly specialized. Functionally, the body is divided into head, thorax and abdomen. The **exoskeleton** is of **chitin** (a nitrogenous polysaccharide) and **appendages** are **jointed.** Arthropods have a complex eye structure and highly developed organ systems. The tremendous diversity (approximately 800,000) species and vast numbers of individuals are indicative of their adaptability or "biological success." **Digestive systems** are **complete** and **circulatory systems** are usually of the **open type.** The **respiratory system varies** with the environment in which the organism lives — aquatic or terrestrial. **Excretory systems vary** widely, but all are relatively efficient.

Class Chilopoda (Centipede) and Class Diplopoda (Millipede)

1. Observe the demonstration specimens. Is there a distinct head?_________ Is the body rounded or flattened? _____________

Segmented or non-segmented? _________________

2. What type of symmetry is present? ________________________

3. Note the rigid exoskeleton. How can these organisms grow when encased in a rigid casing of this type? ____

Class Arachnida

1. Study examples of this class, which differ from other arthropods in the absence of compound eyes, antennae and true jaws.

2. Is there a separate head or is it fused with the thorax? ________________________

There are_____ pairs of walking legs.

Class Crustacea

Most crustaceans are marine although some inhabit inland waters and a few, such as sow bugs, are able to live in moist terrestrial environments. Embryological studies show that the body is made up of a series of metameres or segments, but in adult animals only the abdomen appears segmented. Various crustaceans are on demonstration.

1. Each pair of students should get a crayfish from the container and place it in a dissecting tray.

2. As you proceed with the dissection, look for characteristics common to most members of this phylum and compare the specimen with **Figure 20.2** and with charts and models on display.

3. **Phylum Characteristics**

Jointed chitinous exoskeleton, jointed legs, segmentation of the body.

a. As you study the specimen and locate structures, label **Figure 20.2.** On the crayfish note that head and thorax are fused into a **cephalothorax.** The saddle-like covering over the cephalothorax is the **carapace.** A transverse groove separates the fused head from the thoracic region. Laterally the carapace covers the gills. The **rostrum,** the anterior pointed extension of the head, has the eyes on either side of it. The abdomen consists of several segments, terminated by a **telson** (an extension of the last segment). Spreading the telson and rapidly drawing it forward under the body causes the crayfish to dart backward. Each abdominal segment contains several plates.

b. The nineteen pairs of **jointed appendages** are variously modified. On the head are appendages used for sensory purposes (touch and taste), equilibrium, biting and handling food. On the thorax are appendages for touch, taste, food handling, offense and defense, and walking. Abdominal appendages are modified for swimming and reproduction. Start with the uropod and work anteriorly to carefully remove each appendage. Observe the structure, check it against its stated function (see lab chart) and note the increasing degree of structural specialization as you proceed anteriorly.

Muscles are attached to the exoskeleton. As you proceed with the dissection, watch for muscle bundles and note how they are arranged.

Sensory organs are well developed. Note eyes and sensory functions of some appendages.

Respiratory system. To expose the gills make a cut with scissors following line 1 on **Figure 20.3,** starting at the end marked with an arrow. Remove a gill and observe its feathery appearance. Cut across the gill and look for blood vessels. Why would gills need a good blood supply? ______________________________________ Continue the incision along the dorsal side following line 2 on the diagram and make another cut along line 3. Remove the carapace sections. Look for additional abdominal internal structures, removing more exoskeleton as needed. Verify your identification by comparing the specimen with charts and models.

Circulatory system. Look along the dorsal midline for the heart in its pericardial sinus. Do arthropods have an open or a closed circulatory system? ______________________

Excretory system. Excretion is partly conducted by **green glands** which have an opening near the antennae. These glands are flesh-colored in preserved specimens.

Digestive system is a tube-within-a-tube, two-opening type. Do they have a coelom?________

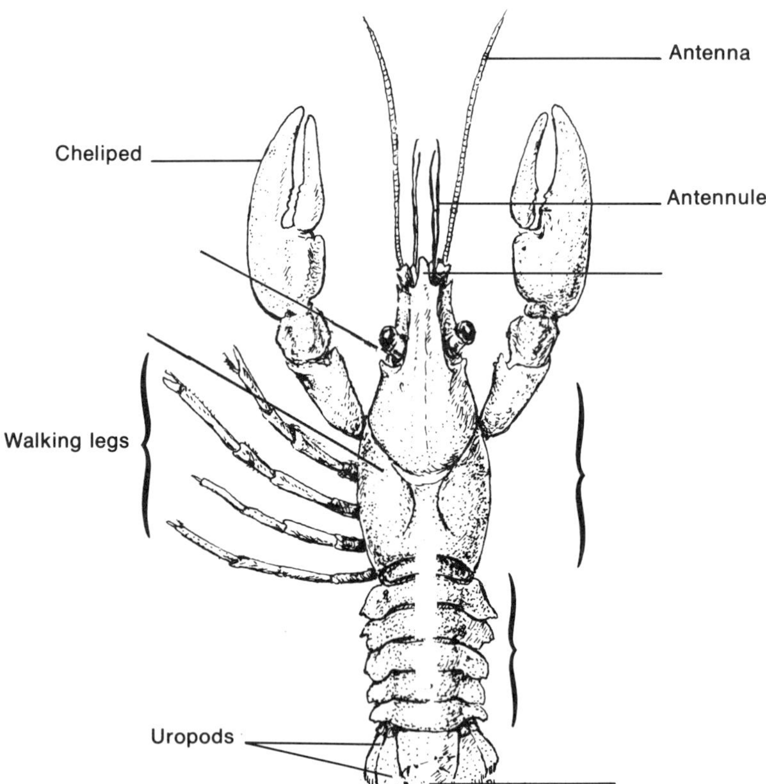

Figure 20.2. Crayfish, dorsal view, walking legs on right side have been omitted for easier labeling.

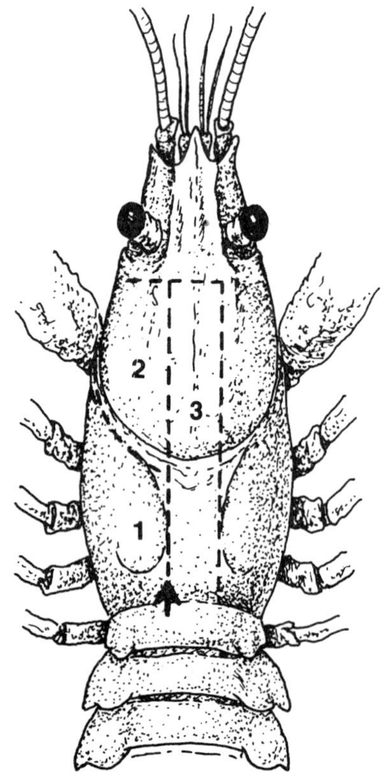

Figure 20.3. Dissection guide for the crayfish.

Class Insecta

Insects are the principal invertebrates able to live in a dry environment and are the only ones capable of flight. Terrestrial life is possible because of a chitinous exoskeleton (protects and prevents desiccation) and the system of tracheal tubes (enabling them to breathe air). The ability to fly allows greater dispersal and increased chances of obtaining food.

1. Obtain a specimen of a grasshopper and label **Figure 20.4** as you examine the specimen.
2. **Head** — consists of six fused segments or somites. How many pairs of **antennae** are there?__________Are they jointed?__________Locate the two large **compound eyes** and the **ocelli. Mouthparts** are of the chewing type and include a number of structures.
3. **Thorax** — consists of three somites. To what region are the jointed legs and wings attached? ____________________ The **anterior wings** are thickened, giving protection for the larger posterior pair of **flight wings.** Wings are derived from the cuticle and are strengthened by the branching vein structure. The outer exoskeleton consists largely of chitin (secreted by the epidermis). Pigment in and under the cuticle provides a protective coloration pattern by which the grasshopper blends with its surroundings.
4. **Abdomen** — consists of eleven somites, with the terminal ones modified for reproduction. In the male the terminal segment is blunt. In the female the last segment has four sharp conical prongs or **ovipositors** for egg laying. Is your specimen a male or a female?__________Along the lower sides of the abdomen are ten small openings, the **spiracles,** which connect to the system of trachea or air tubes. On the lateral surface of the first abdominal segment is a large circular area covered by a thin membrane. This structure is the auditory organ **(tympanum).**

Other insect orders are on demonstration. Note particularly the differences in wing number and morphology.

Two unidentified invertebrates are on display. Can you identify them?

Unknown #1 Phylum ________________ Reason: ________________

Unknown #2 Phylum ________________ Reason: ________________

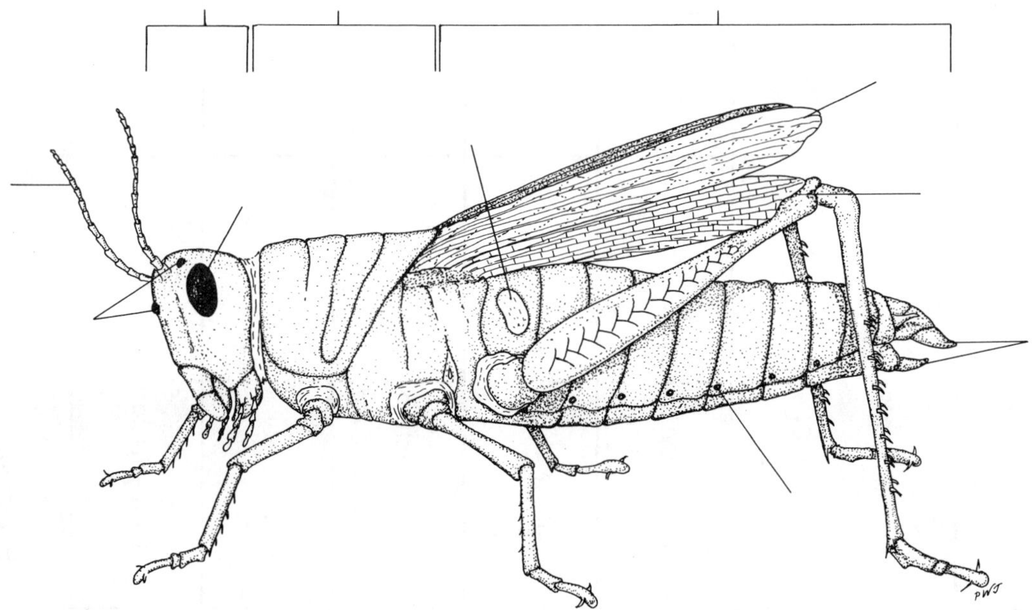

Figure 20.4. External anatomy of the grasshopper.

A COMPARATIVE STUDY OF SOME OF THE INVERTEBRATE PHYLA STUDIED SO FAR

	Phyla					
	Coelenterates	Platyhelminthes	Aschelminthes	Molluscs	Annelids	Arthropods
Principal Characteristics						
Type of Symmetry						
Habitat						
Reproduction						
Body Cavity						
Segmentation						
Examples						

LABORATORY 21

ECHINODERMS AND CHORDATES

Animals with complete digestive tracts may be grouped into two major categories, protostomes and deuterostomes. To understand the reasoning behind these designations it is necessary to study embryological development. The embryonic gut or archenteron has a single opening to the outside, the blastopore. The gut will eventually have two openings, a mouth and an anus. Does the blastopore become the mouth or the anus? In molluscs, nematodes and arthropods the blastopore becomes the mouth and a new opening forms the anus. This is the reason for designating this group the **protostomes** (proto = first, stome = mouth). In echinoderms and chordates the blastopore becomes the anus and the mouth forms from a new opening. Thus the use of the designation **deuterostome** (late or second mouth) for these phyla. Both groups have "one-way traffic" through the gut, but differences in embryonic origin of the mouth openings suggest two major evolutionary pathways in the animal kingdom. The two groups also differ in other embryological features such as type of cleavage and origin of mesoderm.

PHYLUM ECHINODERMATA

Members of this phylum are abundant in the world's oceans. Their "spiny skin" gives them a great deal of protection and they have few enemies. Adult echinoderms are **radially symmetrical,** but **larval forms** (bipinnaria) **are bilaterally symmetrical.** Echinoderms are the first animals you have studied which possess an **endoskeleton,** a hard, internal, mesodermal skeleton. Organ systems are fairly complex. The unique **water vascular system** is characteristic of the group.

1. The starfish *Asterias,* typical of the class Asteroidea, is found along rocky seashores wherever molluscs, its principal food, are abundant. Starfish range in size from one-half inch to eighteen inches in diameter. As you study this animal keep in mind the phylum characteristics.

2. Obtain a specimen of *Asterias* for each table. Can you locate a head?__________What type of symmetry is evident? ____________________

3. The body is divided into a central **disc** from which radiate arms or **rays.** (As you locate the indicated structures, label **Figure 21.1.)** Located at the tip of each ray is a **light sensitive spot.**

4. Two surfaces may be distinguished. The **aboral** or dorsal surface is more rounded and more spiny in appearance. The disc on this side has a small circular body, the **madreporite,** located slightly off center and between two arms. The madreporite is the entrance to the water vascular system.

5. Examine one of the rays under water. Note the blunt calcareous **spines** projecting from the surface. Between the spines are some pincer-shaped structures, the **pedicellariae,** thought to keep the surface free from debris. A hand lens or a dissecting scope will help in these observations. Small thin-walled sacs, the **branchial papillae,** also found between the spines, function in respiration.

6. The **oral** or ventral surface is slightly flattened and contains the centrally located **mouth.** A cluster of **oral spines** project over the mouth. **Ambulacral grooves** radiate out from the mouth. Within the grooves are a number of finger-like projections, the **tube feet.** How are these arranged?

____

Look particularly at the structural arrangement of the tip of these tube feet. Do these remind you of an organism studied last week? ______________________

What are two functions of the tube feet? ______________________

7. Cut one arm off of your specimen and study the cross section of the stump. Look for the arrangement of the plates with their fixed dorsal spines and movable ventral spines along the ambulacral groove. Is this an endo- or an exoskeleton? ______________________

8. Cut the tip off of another arm. Make two lateral cuts extending toward the disc and join the cuts when you get to the disc. Carefully lift off the skeletal material. Make a circular cut to remove the aboral disc. Try to leave the madreporite attached to the specimen.

9. Note that most of the coelom in each arm is occupied by two highly branched **digestive glands** (**Figure 21.2**), which connect by ducts to the centrally located **stomach.** The mouth empties into the stomach through a short esophagus. If your specimen was obtained during the breeding season, the other major structures in the arm are the **gonads** or reproductive organs. The **water vascular system** consists of a series of tubes which conduct water from the madreporite into bulk-like structures **(ampullae).** Contraction and expansion of ampullae combined with the suction discs at the end of the tube feet enable the starfish to move.

Examine and note the characteristics of other echinoderm classes on demonstration.

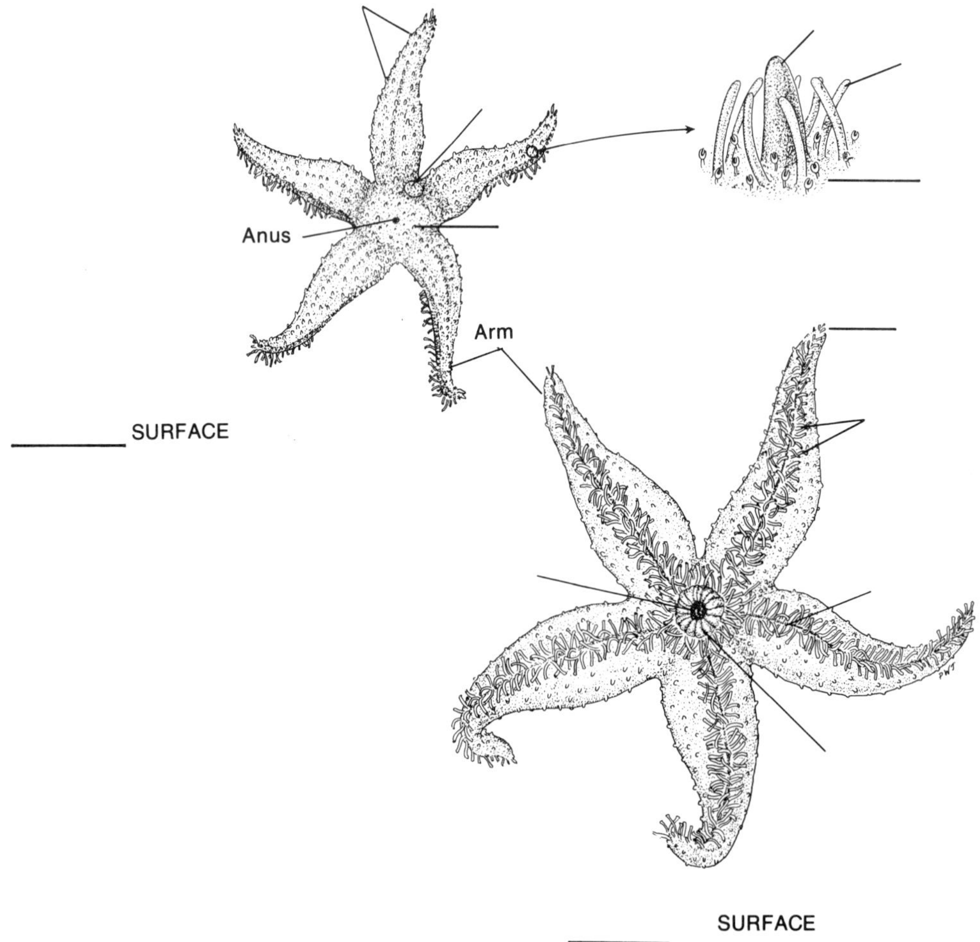

Figure 21.1. External anatomy of the starfish.

Figure 21.2. Internal anatomy of the starfish.

PHYLUM CHORDATA

Chordates share the following characteristics.

1. **Notochord** — a semisolid, cartilage-like support rod which runs the length of the body. It is found in adults of the primitive chordates, but only in embryos of the higher chordates. It is dorsal to the gut and ventral to the nerve cord, and is the precursor of the endoskeleton which will develop.
2. **Pharyngeal gill slits** — perforations in the wall of the pharynx. Water moves into the mouth and out through the clefts. These slits are functional in aquatic types, but in the higher terrestrial forms appear only during embryonic life and are transformed later into other structures useful to air-breathing organisms.
3. **Dorsal, hollow nerve cord** — Animals studied previously had a ventral nerve cord. In chordates it is dorsal to the gut and on the same side as the brain.

Chordate Subphyla

One subphylum contains the rather unusual appearing animals known as the **tunicates.** Observe the demonstration specimens. The adult sessile form has little resemblance to other members of the phylum. The larval tunicate, however, is the major reason for the inclusion of tunicates in this phylum.

Another subphylum contains some small marine animals known as *Amphioxus* or lancelets. Some of these organisms are available.

1. Can you see a notochord?__________ In what other exercise did you observe a notochord?

2. Locate the gill slits and the mouth. Is segmentation seen? __________

The **Subphylum Vertebrata** contains most of the organisms (including yourself) that we usually think of as "animals."

1. Typical examples are on display, along with the characteristics of each class. Study these organisms, noting characteristics that you can determine externally and those which would require dissection, knowledge of their embryonic development, etc.
2. **Vertebrate Skulls and Teeth.** As you study vertebrates, it is possible to note a number of adaptations to various modes of life. A series of mammalian skulls are on display which mirror the varied eating habits of animals. By observing the **zygomatic arch** (your cheek bone, for example), the **dentary** (lower jaw) and the **diastema** (space separating the incisors from the cheek teeth), one can learn much about the eating habits of different mammal groups.

INSECTIVORES (insect-eating). Teeth have low **crowns** and **cusps** are sharply pointed. **Canines** look much like **incisors (Figure 21.3).** The brain case is small and zygomatic arch reduced or missing. A mole's **dental formula** (the number of each kind of tooth in each half-jaw beginning at the anterior midline) is

Upper Jaw	3 incisors — 1 canine — 4 premolars — 3 molars	× 2 = 44
Lower Jaw	3 incisors — 1 canine — 4 premolars — 3 molars	

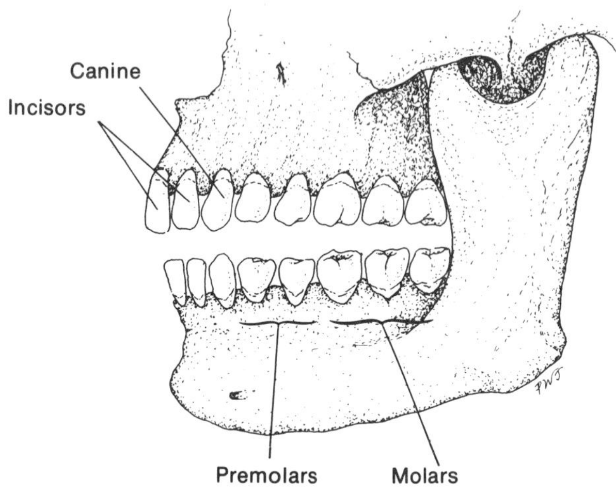

Figure 21.3. Tooth types in the human. Shape, number and types vary in different animal types, reflecting the diet. Humans are **omnivores** (eating both plants and animals) and no tooth type predominates.

CARNIVORES (flesh-eating). Incisors and canines are sharply pointed. What purpose would this serve? ______

Premolars and **molars** are flat, but edges are sharp for cutting. Observe the way the upper and lower jaw hinge together, allowing little lateral movement of the lower jaw. Teeth of the upper and lower jaws fit closely as jaws are closed, giving a shearing action. What might be the function of the prominent zygomatic arch? ______

______

A cat's dental formula is

Upper Jaw	3 incisors — 1 canine — 4 premolars — 2 molars	× 2 = 40
Lower Jaw	3 incisors — 1 canine — 4 premolars — 2 molars	

RODENTS (gnawing). Note the prominent incisors. The outer side has hard enamel, while the inner surface is softer cement and wears away quickly. Is this an advantage? ______

Why? ______

Are canines present? ______ Note the extremely wide diastema and the well developed zygomatic arch. The hinge is so arranged that the jaw slides backward and forward, rather than side to side.

HERBIVORES (plant-eating). The skull of the horse was used previously to illustrate evolution. It can also be used as an example of an herbivore. How does the upper jaw dentition differ from the lower?

______

A typical herbivore is $\frac{0—0—3—3}{4—0—3—3}$

How does the lower jaw move? (side to side, or backward and forward) Is this type of movement advantageous to an herbivore? ____________

Refer to Exercise 16 for additional information on tooth surface and characteristics typical of herbivores.

UNKNOWN SKULLS (Demonstration). Can you identify these skulls as fitting into one of the four categories just discussed?

Unknown #1 Type ________________ Reasons: ________________

Unknown #2 Type ________________ Reasons: ________________

3. Vertebrates will be discussed further in the museum exhibit. There will be a short introductory talk and then specimens will be available to demonstrate various aspects — such as type of food related to beak shape, type of feet related to habitat and/or food getting habits, etc. Most of this is self-guided with explanatory cards and posters. **You will need to take notes on what you see.**

4. Checking off the various characteristics on the summary sheet will help you to categorize and review the different chordate groups.

SUMMARY CHART OF CHORDATES

		Tunicates	Lancelet	Jawless Fishes	Cartilaginous Fishes	Bony Fishes	Amphibians	Reptiles	Birds	Mammals
Dorsal hollow nerve cord										
Notochord	in embryo and adult									
	embryo only									
Gills	in embryo and adult									
	embryo or larva only									
Jaws										
Two sets of paired fins										
Tetrapod limbs										
Feathers										
Wings										
Hair										
Mammary glands										
Skeleton	bony									
	cartilaginous									
Homeothermic										

ECHINODERMS AND CHORDATES CROSSWORD PUZZLE

Across

1. Organism whose blastopore becomes the anus
3. A chordate dorsal nerve cord is____.
4. type of symmetry seen in an adult starfish
6. opening to the water vascular system:____porite
8. All organisms studied in the previous exercise were____stomes.
11. Starfish larvae have____symmetry.
12. The Amphioxus or____let has prominent somites.
14. characterized by a dorsal nerve cord, gill slits and a notochord
16. perforations in the wall of the pharynx:____slits
17. The madreporite is located in the____on the aboral side.
18. bipinnaria____(plural)
20. On the starfish, blunt calcareous____project from the surface. (singular)
21. A wolf is____ivorous.
24. Reduced or missing zygomatic arches and canines like incisors are typical of____ivores.
26. Starfish move by means of____feet.
27. The starfish mouth is on the____side.
28. Sharply pointed incisors and____are associated with carnivores. (singular)
29. An extra wide____is found in rats and mice.
30. The horse____can be used to tell about its eating habits.

Down

1. Chordate notochords are on the____side.
2. Echinoderm dissected in lab
3. Side to side movement of the jaw is characteristic of____ivores.
5. Both groups studied in the lab today have____ternal or endoskeletons.
6. A group of oral spines surround the starfish____.
7. Echinoderms and primitive chordates live in____.
9. Echinoderms have a unique water____system.
10. Chordates have a dorsal____cord.
11. The____becomes the mouth in molluscs and annelids.
13. The madreporite is on the____side.
14. Annelids have ventral nerve____.
15. "Spiny skin" organisms are____.
19. primitive sessile chordate (backward)
22. support structure of the lancelet:____chord
23. grinding teeth
25. have hair and secrete milk:____ls
26. canines, incisors, molars

(see page 209 for correct answers.)

CROSSWORD PUZZLE ANSWERS

CHEMICAL NATURE OF LIVING ORGANISMS

ACROSS

2. neutron 4. taf 7. ucose 9. bon 11. periodic 14. atomic number 16. spring 17. neg 18. noi 19. OH 20. ron 22. atoms 24. oxygen 26. cellul 29. dipil 30. synthe 32. share 34. ring 35. adenine 38. carbohydrate 40. glyc 41. ADP 42. mon 43. reduction 45. proteins 46. eotides

DOWN

1. covalent 3. elements 5. formic 6. iso 8. energy 10. ino 11. proton 12. disaccharide 13. carb 15. methy 19. oxyl 20. recylg 21. glyc 22. ATP 23. oxida 25. igh 27. lysis 28. rib 31. weight 32. single 33. remylop 36. acid 37. sylop 39. hate 42. mon 44. edi

TRANSPORT SYSTEMS AND CIRCULATION

ACROSS

1. doolb 3. aorta 9. reptile 10. lunar 11. owt 14. tri 15. veins 16. circulatory 18. red 20. sinus 21. hemoglobin 22. valve 24. pulse 26. fis 27. bi 28. cyte 29. erythro 30. mamma 31. cuspid

DOWN

2. bird 4. rate 5. artery 6. open 7. closed 8. gills 10. leuco 12. pulmonary 13. plasma 14. trium 15. venosus 17. thr 19. chambers 20. semi 22. ventr 23. llari 25. effac 26. fo

PATTERNS OF INHERITANCE

ACROSS

2. egg 5. dihybrid 8. gen 11. type 14. homozygous 18. hetero 19. loci 20. auto 25. allele 26. filial 27. XX 29. body 30. somes 33. gene 35. zygotes 37. LE 39. linked 40. sperm 41. multiple 42. zygous 43. parental 44. cross

DOWN

1. XY 3. gg 4. geno 5. dominant 6. homo 7. ratio 9. pheno 10. Punnett square 12. philia 13. Ee 15. or 16. mono 17. genetics 21. trait 22. asex 23. blood 24. first 27. xy 28 recessive 31. spring 32. X^hY 33. ge 34. groups 36. gametes 38 ERER

DEVELOPMENT

ACROSS

1. pigment 3. chalazae 5. air 6. ecto 8. meso 11. eggs 12. zygote 13. furrows 15. disc 17. holo 18. meri 20. embryos 24. anus 25. mor 27. coel 28. shell 30. hook 31. vegetal 32. in 34. epicot 36. organo 37. lip 38. tula 39. germ 40. cleavage 41. leaf

DOWN

1. pore 2. noto 3. cells 4. all 5. amn 7. grow 8. morpho 9. shoot 10. mero 11. endo 12. zone 14. seeds 16. somite 17. hypo 18. membranes 19. root 21. yolk 22. sacs 23. archen 25. mitosis 26. placenta 29. gastrula 31. vitel 33. ova 34. elong 35. cotyl

VASCULAR PLANTS

ACROSS

1. tracheophyta 8. needle 9. troppus 11. stigma 14. psilopsida 17. pteropsida 19. tiurf 20. entire 21. lobe 23. ferns 24. ovary 25. axillary 27. simple 30. cot 32. embryo 34. cycad 35. opposi 36. *setum* 37. bud 38. sphenop 39. seed

DOWN

1. transport 2. cone 3. poll 4. angiosperms 5. petal 6. vascular 7. mono 10. pisti 12. di 13. conifers 15. sepal 16. anthe 18. petiole 22. blade 23. flowers 25. alternate 26. food 27. spermop 28. mar 29. lycop 30. compound 31. taoc 33. rush

ECHINODERMS AND CHORDATES

ACROSS

1. deuterostome 3. hollow 4. radial 6. madre 8. proto 11. bilateral 12. lance 14. chordate 16. gill 17. disc 18. larvae 20. spine 21. carn 24. insect 26. tube 27. oral 28. canine 29. diastema 30. skull

DOWN

1. dorsal 2. starfish 3. herb 5. in 6. mouth 7. water 9. vascular 10. nerve 11. blastopore 13. aboral 14. cords 15. echinoderms 19. etacinut 22. noto 23. molars 25. mamma 26. teeth

REVIEW MATERIAL FOR EXAMS

All boldface terms and questions are fair game for exams. Other important material is listed below.

Laboratory 1. Scientific Measurement, the Scientific Method and the Chemical Nature of Life

1. Use and interconvert metric system units.
2. Know the parts of the scientific method.
3. Know tests used, substances tested for, and interpretation of results.

Laboratory 2. The Chemical Structure of Living Organisms

1. Use of periodic tables and atomic structure.
2. Types of bonds.
3. Oxidation-reduction — what happens; recognition of this process on models.
4. Recognition of model molecules made in the lab (both intact molecules as well as ions and groups.)
5. Synthesis and dehydration reactions involving 4 major biological molecule categories; components of the biological molecules.

Laboratory 3. Enzymes — Biological Catalysts

1. Characteristics of enzymes and factors or conditions which alter their activity.
2. Tests used to show (a) production by organisms, (b) chemical composition, (c) end-product, (d) effect of inhibitors, temperature and concentrations on activity.

Laboratory 4. Microscope Use and Physical Processes Important to Life

1. Recognition of microscope parts and their functions.
2. Information which can be obtained with a microscope.
3. Differentiate between Brownian movement and true motility.
4. Know what happens in different types of diffusion situations.

Laboratory 5. The Cell as a Unit of Structure

1. Eucaryotic vs. procaryotic cell structure.
2. Recognize organelles and structures of plant and animal cells — on models, through the microscope and on diagrams and photographs.
3. Know functions of cell structures.
4. Know differences between plant and animal cells.
5. Recognize morphological types of bacteria.

Laboratory 6. The Cell as a Unit of Function

1. Typical cell shapes, functions and arrangements.
2. Recognition of structures in plant and animal tissues studied.
3. Know functions of tissues studied.
4. Recognize tissues in an organ (frog intestine).

Laboratory 7. Cellular Metabolism and Nutrition

1. Know respiration equation and tests used to show CO_2 end product.
2. Autotrophic nutrition — equation, important factors and how tested for.
3. Heterotrophic nutrition — three classifications; required materials for these organisms; differences observed in holozoic nutrition in paramecium, water flea, earthworm and frog; structures in cross section of frog and earthworm intestine and their functions.

Laboratory 8. Transport Systems and Circulation

1. Types of circulatory and exchange systems and examples of organisms which use them.
2. Know function of and be able to identify heart structures on vertebrate models and specimens.
3. Identify heart model as belonging to fish, bird, mammal, reptile or amphibian.
4. Know effect of exertion and caffeine on human heart rate.
5. Know functions of erythrocytes and leucocytes.

Laboratory 9. Water Balance and Excretion of Waste Products

1. Explain transpiration results (relating to structure and photosynthesis).
2. Recognize leaf structures and know their function.
3. Recognize models of protein breakdown end-products and know advantages and disadvantages of each.
4. Interpret tests which used Nessler's reagent.
5. Identify vertebrate kidney structures.

Laboratory 10. Receptor — Conductor — Effector

1. Know receptor categories and how these were tested for.
2. Recognize taste bud structures.
3. Know conductor structure.
4. Trace evolutionary advancements in primitive nervous systems on up through higher animals.
5. Advancements associated with nervous systems of bilaterally symmetrical animals.
6. Recognize areas in vertebrate brains and relate relative importance (size) of these to habitat or activities of the organisms to which the brain belongs.

Laboratory 11. Organ Systems in an Invertebrate and a Vertebrate

1. Recognize earthworm structures and know their functions.
2. Recognize frog structures.
3. Know organs which function in the various organ systems observed in both of these animals.

Laboratory 12. Mitosis and Meiosis

1. Sequence of phases.
2. Recognition of phases from (a) microscope slide, (b) photograph of microscope slide, (c) plaster model and (d) pipecleaner model.
3. Differences between mitosis in plant and animal cells.
4. Identification of structures visible in a dividing cell.
5. End result of mitosis and meiosis (number of cells, chromosome numbers).

Laboratory 13. Patterns of Inheritance

1. Use given information to determine possible phenotype, genotype and types of gametes. (Therefore, must know how to work monohybrid and dihybrid crosses and crosses involving sex chromosomes and blood groups).
2. Antigens and antibodies of the different blood groups, and reaction expected when different ones are combined.

Laboratory 14. DNA Structure and Protein Synthesis

1. Components of DNA and RNA.
2. Differentiate between written structures or models of purines and pyrimidines (not between A and G or C and T), ribose and deoxyribose.
3. Bonding characteristics of nitrogen bases.
4. How information is coded.
5. Sequence and events of protein synthesis.
6. Types of mutations (of DNA) and results of each.
7. Why you looked at smooth and wrinkled peas, corn, flies, etc.

Laboratory 15. Animal and Plant Development

1. Sequence and definitions of four major periods of development.
2. Recognition of stages (and sequence of) in starfish and frog slides and models.
3. Effect of yolk quantity on developmental events.
4. Identification, function and importance of egg membranes, somites and notochord in chick embryos.
5. Types of cleavage.
6. Identification and function of structures in seed, seedling and root tip.
7. What each part of the seed will form in the seedling.

Laboratory 16. Adaptation and Evolution

1. Recognition of structures (and their functions) in the three leaf types.
2. Insect adaptations (coloration and mimicry). Be able to identify which of these adaptations is exhibited by a specimen.
3. *Clonorchis sinensis* — structures and their functions; how its structures relate to its parasitic life style.
4. Horse — trends seen in skull and leg as horse evolved; correlation of leg and tooth types with habitat; interpretation of graphs (e.g., What is illustrated by a vertical line? by a horizontal line? etc.)

Laboratory 17. Monera, Protista and Lower Plants

1. Differentiate between procaryotic and eucaryotic cells.
2. Division and Phylum names and common names for the classes.
3. Place of each group in a food chain (producer, consumer, decomposer).
4. Monera — characteristics; examples of good and bad activities; recognize morphological types.
5. Protista — kingdom characteristics; division and phylum characteristics; recognition of examples, structures.
6. Plantae — kingdom characteristics; why algae are grouped together; examples and characteristics of the algal divisions.

Laboratory 18. Bryophytes (Non-Vascular) and Tracheophytes (Vascular Plants)

1. Plantae — kingdom characteristics; differentiate verbally and visually between the classes and the n and 2n stages of bryophytes; trends seen.
2. Why tracheophytes are the dominant terrestrial plant group.
3. Characteristics and recognitions of Psilopsida, Lycopsida, Sphenopsida and Pteropsida.
4. Advances seen as one progresses from Psilopsida to Pteropsida.
5. Seed components.
6. Cycads and conifers — characteristics and examples.
7. Angiosperms — two important features; flower parts; distinguish between monocots and dicots.
8. Leaf types, parts and arrangements.
9. Evolutionary sequence of plants.

Laboratory 19. A study of Fungi, Lichens and Several Animal Phyla from Sponges to Flatworms

1. Recognize symmetry types.
2. Know phylum characteristics.
3. Examples and characteristics of the fungi studied; functions of lichen components.
4. Sponges — basis for separating into classes, recognition of examples.
5. Coelenterata — polyp vs medusae, hydra, obelia and aurelia structures and recognize other phylum members.
6. Plathyelminthes — advances seen in this phylum (compared to previous phyla); specializations for a parasitic mode of life; structures of planaria and tapeworms.

Laboratory 20. Some More Advanced Invertebrates

1. Aschelminthes — advances illustrated.
2. Mollusca — unique characteristics, classes, examples.
3. Annelida — characteristics of the group; examples.
4. Arthropods — recognition of various classes as belonging to this phylum; grasshopper structures.

Laboratory 21. Echinoderms and Chordates

1. Difference between protostomes and deuterostomes.
2. Echinoderms — starfish structure and functions; recognition of other echinoderms as belonging to this phylum.
3. Chordates — recognize tunicates and lancelets and know why included in this group. Vertebrates — classes and their characteristics and examples of each; skulls and teeth and relation to eating habits.
4. Museum tour — types of, and functions of museums.
   Amphibians — locomotion, feeding, defense.
   Reptiles — why successful; locomotion, feeding adaptations.
   Fishes — feeding, locomotion, reproductive adaptations.
   Birds — adaptations illustrated by feet, bills, shape and color of eggs, flight.
   Mammals — three reproductive types; defense and foot adaptations.